I0491743

Hydraulic Pipes, Tubes, and Hoses

(In the SI Units)

Joji Parambath

Hydraulic Pipes, Tubes, and Hoses
(In the SI units)

Copyright © 2026 Joji Parambath

ISBN: 9798653889455

https://jojibooks.com

First Edition 2020
Revised Edition 2026

Disclaimer of Liability

The contents of this book have been checked for accuracy. Since deviations cannot be precluded entirely, we cannot guarantee full agreement. Only qualified personnel should be permitted to install and service hydraulic equipment. Qualified persons are defined as persons who are authorised to commission, ground, and tag circuits, equipment, and systems following established safety practices and standards.

Table of Contents

Chapter	Description	Page No
	Preface	v
1	Fluid Conductors – Introduction and Terms & Definitions	1
2	Hydraulic Pipes and Fittings	13
3	Hydraulic Tubes and Fittings	33
4	Hydraulic Hoses and Fittings	41
5	Design of Hydraulic Piping Systems	54
6	Installation, Routing & Maintenance of Fluid Conductors	67
7	Objective Type Questions	78
8	Review Questions	79
9	Numerical Problems	82
10	Refcrences	83

PREFACE

Fluid conductors interconnect components of a hydraulic system for the safe and leak-free transmission of high-pressure hydraulic fluid throughout the system. As hydraulic systems become increasingly complex, operating at higher temperatures and in limited spaces, not only must the fluid conductors withstand these adverse conditions but also handle high working pressures, peak surge pressures, and peak flow rates. A wide range of hydraulic applications requires various conductor types to meet varying operating requirements and conditions.

This book presents information on the constructional features, performance specifications, and other details of pipes, tubing, and hoses, along with their fittings. Next, the topics are logically arranged for a simple-to-complex level progression of the subject matter. The book uses the SI system of units.

Many other fluid power topics are covered in other textbooks in the same author's fluid power educational series. A list of all the textbooks is provided at the end of the book. Please also see the details at https://jojibooks.com.

Enjoy reading the book.
Your feedback is most welcome.

JOJI Parambath

Chapter 1 | Fluid Conductors – Introduction and Terms & Definitions

In a conventional hydraulic system, various components are connected by a network of conductors. This means the conductor system is a network of conductors that connects to system components via fittings, enabling effective fluid delivery throughout the system. Pipes, tubing, and hoses are the three basic types of fluid conductors used in hydraulic systems.

A conductor is a pressure-tight vessel that conveys a sufficient quantity of pressurized fluid through it in a leak-free manner. It must have smooth interior surfaces to reduce friction and turbulence in the fluid and sufficient wall thickness to withstand the system's high operating and shock pressures. It must also be capable of withstanding high system and ambient temperatures and be compatible with the fluid used.

The critical considerations for selecting fluid conductors include their construction, sizing, installation, routing, and applicable standards.

Terms and Definitions, Fluid Conductor

The fluid power industry uses numerous conductor-related terms to specify the performance levels of fluid power systems.

For example, a fluid-conducting pipe is specified by its diameter and wall thickness.

The following sections provide brief definitions of commonly used terms for fluid conductors.

Diametrical Size

The diametrical size of a conductor is specified by its inside diameter, outside diameter, or nominal size (Figure 1.1). These specifications are important because using the correct conductor diameters can prevent high flow velocities and excessive pressure drops.

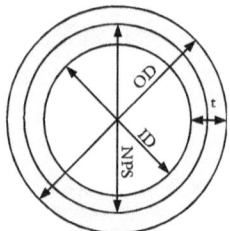

Figure 1.1 | Size specifications of a fluid conductor

Inside Diameter, D_i

It is the smallest cross-sectional diameter of a conductor. It decides the cross-sectional area for the fluid flow. A smaller conductor area results in higher friction, higher pressure drop, greater heat generation, and greater energy loss.

Outside Diameter, D_o

It is the largest cross-sectional diameter of a conductor. It dictates the size of conductor fittings and is also critical for determining routing layouts through confined spaces.

Nominal Size

In ANSI and SAE standards, for example, pipe size is specified by nominal pipe size (NPS), and in the SI system, by Nominal Diameter (DN). A detailed explanation of this topic is given in Chapter 2 under the heading 'Nominal Pipe Size and Diameter Nominal'.

Wall Thickness, t

The wall thickness of a pipe or tube determines the maximum pressure it can withstand. It is expressed in metric units or as a schedule number, as in ANSI and SAE standards. It is given by:

$$\text{Wall thickness, } t = (D_o - D_i) / 2$$

For a hose, the wall thickness indicates its reinforcement strength and pressure-handling capability.

Schedule Number

The schedule numbers vary from 5 to 160 in a graded manner. Altogether, there are eleven different schedule numbers. They are: 5, 10, 20, 30, 40, 60, 80, 100, 120, 140, and 160.

A higher schedule number for a given pipe size indicates a thicker wall. For a given pipe size, the outer diameter remains constant, while the inside diameter decreases as the schedule number increases.

For example, Table 1.1 lists the wall thicknesses for schedule numbers 40, 80, and 160 for a pipe with a nominal size of ½. The wall thickness increases with the schedule number. This topic is further elaborated in Chapter 2 under 'Wall Thickness – Schedule Numbers'.

Table 1.1 | Wall thicknesses of pipes

Nominal Pipe Size	Outside Diameter	Wall thickness		
		Schedule 40	Schedule 80	Schedule 160
	inch	inch	inch	inch
½	0.500	0.840	0.109	0.147	0.188

Example 1.1

Determine the inside diameter of a pipe with an OD of 48.3 mm and a wall thickness of 2.3 mm.

Solution

Pipe OD	= 48.3 mm
Wall thickness	= 2.3 mm
Inside diameter of the pipe	= 48.3 – 2 x 2.3 mm
	= 43.7 mm

Hoop Stress

Hoop stress in a pipe or tube is the circumferential stress acting on the wall of the conductor that tends to split it. It is the maximum pressure the conductor material can withstand before failure. Consider a thin-walled conductor with an inside diameter (Di), wall thickness (t, where t < 0.1 x Di), and length (L), as shown in Figure 1.2. The conductor is subjected to operating pressure, P.

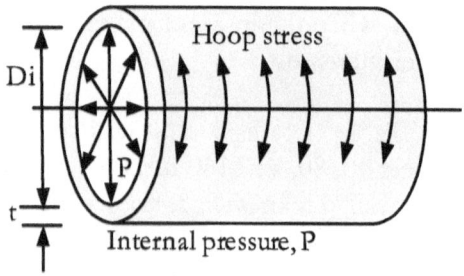

Figure 1.2 | Hoop stress

The operating pressure, acting normal to the inside surface of the pipe, induces a circumferential force in the conductor that tends to split the conductor into two halves. The normal surface area can be taken as the projected area $(D_i \times L)$ of one-half of the pipe.

4

The circumferential force (or burst force) is given by:

Circumferential force = $P \times D_i \times L$

The tensile force, which is trying to resist the splitting of the conductor, acts on the cross-sectional area (tL) of each wall. Therefore, the resistive force is given by:

Resistive force = $2tL \times$ Hoop stress.

Equating the circumferential force to the resistive force, we get,

$P \times D_i \times L = 2tL \times$ Hoop stress

Therefore,

Hoop stress = $P \times D_i \, / \, 2t$

The conductor must have sufficient tensile strength to prevent it from bursting under excessive hoop stress.

Tensile strength: It is the ability of a material to withstand a pulling (tensile) force.

Yield Strength: It is the stress a material can withstand without permanent deformation.

The yield point is the maximum stress a ductile material can endure, beyond which it begins to deform permanently. Tensile strength is the maximum (ultimate) stress a material can withstand before it fractures. For example, the tensile strength of carbon steels typically ranges from 4150 to 7600 bar. (60000 to 110000 psi).

Example 1.2

Determine the hoop stress developed in a pipe of outside diameter 76.1 mm and a wall thickness of 4.2 mm when the pipe is subjected to a pressure of 70 bar.

Solution

Outside diameter of the pipe, D_i = 76.1 mm
Wall thickness of the pipe, t   = 4.2 mm
Pressure, P   = 70 bar

Hoop stress developed   = P x D_i / 2t
                                      = 70 x 76.1 / 2 x 4.2
                                      = 634 bar

Burst Pressure

It is the internal pressure in a fluid-filled conductor that causes it to burst (Figure 1.3). The conductor bursts when the hoop stress exerted on the conductor exceeds the tensile strength (S) of the conductor material.

Barlow's formula is commonly used to predict the burst pressures in ductile thin-wall tubes (t < 0.1 x D_i)

$$\text{Burst pressure (BP)} = 2tS / D_i$$

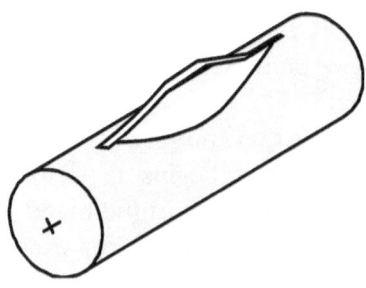

Figure 1.3 | Burst-tested tube

In thick-walled pipes, the tensile stress across the wall thickness is non-uniform. Therefore, the following formula must be used to account for the non-uniform tensile stress.

$$\text{Burst pressure (BP)} = 2tS / (D_i + 1.2t)$$

Working Pressure
The working pressure of a conductor is the maximum pressure to which it can be subjected without damage. It is calculated by dividing the conductor's burst pressure by a safety factor.

$$\text{Working pressure (WP)} = \frac{\text{Burst pressure (BP)}}{\text{Safety factor (SF)}}$$

- A safety factor of 4:1 is used for hydraulic applications where shock and mechanical strain are not considerable.

- A safety factor of 6:1 should be used where considerable shock and mechanical strain are expected.

- A safety factor of 8:1 should be used where severe hydraulic shock and mechanical strain are expected.

Design Pressure
It is the pressure to which each component of a piping system is designed to withstand. It must not be less than the actual pressure at the most severe expected pressure and temperature during service of the piping system.

Maximum Allowable Working Pressure
It is the maximum pressure in a piping system, determined by its weakest component. It is not to exceed its design pressure.

Minimum Bend Radius

It is the smallest radius of the curved section of a conductor (semi-rigid tube or flexible hose) beyond which it should not be bent without flattening, kinking, or wrinkling (Figure 1.4). Bending the conductor beyond its limit causes severe backpressure and internal damage, leading to premature failure.

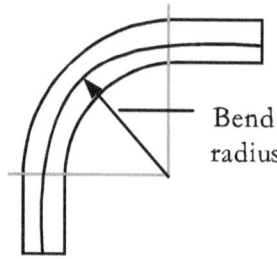

Bend radius

Figure 1.4 | Minimum bend radius

The manufacturer specifies the minimum bend radius for a hydraulic semi-rigid or flexible conductor in accordance with the test standard, which is typically stated in the technical data sheet.

The main factors determining the bending radius of a hydraulic conductor are its inside diameter (ID), wall thickness, and material. A larger ID and thicker wall thickness result in a larger bend radius. During installation, the bending radius of a conductor must exceed its specified minimum bend radius, so sharp bends should be avoided.

Design Temperature

It is the maximum operating temperature for each piping component. It may be taken as the maximum fluid temperature.

Example 1.3

What would be the maximum pressure rating of a pipe of OD of 88.9 mm with a wall thickness of 2.99 mm? The pipe material is stainless steel with a tensile strength of 5500 bar. Assume a safety factor of 6.

Solution

Pipe OD, Do	= 88.9 mm
Wall thickness, t	= 2.99 mm
Tensile strength, S	= 5500 bar
Safety factor, SF	= 6

Pipe ID, Di	= Do – 2 x t
	= 88.9 – 2 x 2.99 = 82.92 mm

Burst pressure	= 2ts / Di
	= 2 x 2.99 x 5500 / 82.92
	=396.65 bar

Pressure rating, pipe	= Burst pressure / Safety factor
	= 396.64 / 6 = 66 bar

Flow Rate

Volumetric flow rate is the volume of oil passing through a cross-section of a fluid conductor per unit of time. It is usually measured in lpm or m^3/s.

Flow Velocity

The velocity of fluid flow is the average speed at which its particles move past a given cross-section. It decides the internal diameter of the pipelines and the type of fluid flow. A smaller conductor increases the average particle speed at a given flow rate, whereas a larger one decreases it, significantly affecting the pressure drop and efficiency.

Flow Rate Vs Velocity of Flow

The rate of flow (Q) of the fluid flowing through a conductor equals the conductor area (A) multiplied by the velocity (v) of the fluid flow. That is,

$$Q = A \cdot v$$

Therefore, at a constant flow rate, the fluid velocity increases as it passes through the narrow section.

Example 1.3

Fluid at the rate of 0.0012 m³/s is flowing through a pipe. Calculate the inside diameter of the pipe so that the fluid particles move at an average velocity of 0.5 m/s.

Solution

Flow rate, Q $\qquad$ = 0.0012 m³/s
Velocity of fluid particles, v $\quad$ = 0.5 m/s

Area of the pipe section, A $\qquad$ = Q/v
$\qquad\qquad\qquad\qquad\qquad\qquad$ = 0.0012 / 0.5 = 0.0024 m²

$$\text{Internal diameter} = \sqrt{\frac{4 \cdot A}{\pi}} = \sqrt{\frac{4 \times 0.0024}{\pi}}$$

$$= 0.0553 \text{ m} = 55.3 \text{ mm}$$

Pipe and Tube Materials

The pipe and tube materials suitable for high-pressure industrial hydraulic service are:
(1) Cold-drawn seamless carbon steel and
(2) Cold-drawn seamless stainless steel.

Stainless steel pipes and tubes are used in applications that require corrosion resistance, such as chemical equipment and marine vessels. Hot-rolled pipes are not recommended for hydraulic services because they leave scale deposits both inside and outside the pipes.

Steel is an alloy of iron and carbon (<2%). Steels are classified into four basic types based on their carbon and other alloying element content. They are carbon steel, alloy steel, stainless steel, and tool steel.

- **Carbon steel** is an iron-carbon alloy. It also contains manganese, silicon, copper, lead, aluminium, cobalt, chrome, and nickel, among others, in varying amounts. Carbon steels are vulnerable to corrosion.

-**Alloy steel** is prepared by adding steel with various metals such as iron, nickel, aluminium, copper, etc. The strength and properties of alloy steel depend on the amount of the elements present in the alloy steel.

-**Stainless steel** has low carbon content but contains chromium alloy, nickel, or molybdenum. It is strong, corrosion-resistant, and capable of withstanding high temperatures.

-**Tool steels** are hard and used to make metal tools.

Steel specifications are issued by organisations such as ISO, AISI, SAE, ASTM, European Standards (EN), and German Standards (DIN), each in its own way. Common materials include carbon steel (SAE 1010, 1020, 1026, 4130, ST37.4, ST52.4) for high-pressure strength and stainless steel (304, 316) for corrosion resistance.

Properties and Standards of Typical Steel Materials

Table 1.2 | Properties and standards of typical steel materials

Pipe material	Properties	Standards
Cold-drawn seamless carbon steel	High-pressure capability, precise dimensions/ shape, clean inside surface with no scale, excellent scaling surface after roll flaring	DIN EN 10305-4 E 355N (St. 52.4 NBK) E 235N (St. 37.4 NBK)
Cold-drawn seamless stainless steel	High-pressure capability, precise dimensions/ shape, excellent scaling surface after roll flaring	DIN EN 10216-5 ASTM A269/ A213 ASTM A312

Tensile Strengths and Yield Strengths of Steels

Table 1.3 | Tensile strengths and yield strengths of steels

Steel type	Tensile strength (N/mm^2)	Yield strength $(N/mm^2$ min)
E235N tubes (St 37.4)	340	235
E355N tubes (St 52.4)	490	355
AISI 316L metric size tubes	485	170
TP 316L schedule size pipes	485	170
DIN 2391 St. 45	570	255
DIN 2391 St. 52	630	355

E – Steel for machine parts
235 – Minimum yield strength in N/mm^2

Chapter 2 | Hydraulic Pipes and Fittings

Pipes are rigid conductors with relatively thick walls used to contain and convey hydraulic fluids. They are highly resistant to bending. It is difficult to shape rigid pipes into the desired configuration. Remember, configuring a piping system is more labour-intensive. Many fittings, such as elbows and tees, are needed when routing a piping system. They are liable to transmit shock and vibration between components. All piping segments should be secured with clamps, preferably damped ones, to absorb shock and vibration and prevent their transmission.

Construction details, Pipes

Pipes have a higher wall thickness but are much cheaper than tubes and hoses. Hence, they are generally employed in applications where there is no restriction on conductor size and cheaper conductor systems are preferred. It is recommended to use pipes with higher tensile strength, where possible, because higher tensile strength allows higher permissible working pressures and thinner walls, reducing overall weight in the pipe and supporting structures. Pipes are usually not used in mobile hydraulic systems.

Cold-drawn seamless carbon steel and austenitic stainless steel pipes are used in hydraulic systems for their strength. With these materials, pipes can be precisely shaped and sized. They maintain good cleanliness, with no scale. In general, carbon steel is used for pipes employed in indoor hydraulic applications. Stainless steel pipes are used in applications that require corrosion resistance, such as chemical equipment and marine vessels. A galvanised pipe is not recommended for use in hydraulic systems.

Advantages and Disadvantages of Pipes

Pipes have many advantages and disadvantages. Some of them are listed below:

- A conductor system with steel pipes is the least expensive way to assemble a hydraulic system with low to medium pressure ratings
- As pipes are made of inflexible material and have a large wall thickness, they are difficult to form into the desired configuration and install
- They cannot withstand high surge pressures

Thermal Expansion of Pipes

Significant temperature variations can occur in some hydraulic systems, especially in marine and offshore applications. Under certain conditions, the temperature can vary widely, for example, from –40°C during winter to +40°C during summer. Variations in temperature cause the pipes to expand. Remember, a pipe length can vary by almost 1 mm per 1-metre length of the pipe when the temperature difference is 80°C.

Basic Requirements of Pipes

The basic requirements of pipes for hydraulic service are:

- Pipes must have sufficient cross-sectional areas to satisfy the flow rate requirements without producing excessive pressure drops
- They must be strong enough to withstand the working pressure, shock pressures, and vibration
- They should have smooth interiors to reduce the friction and flow turbulence
- They must be compatible with the type of fluid used
- They must withstand high operating temperatures
- They must be supported by damped mountings to absorb both shock and vibration

Pipe Size Specifications

A pipe should have a sufficient ID and a smooth inside surface to reduce frictional forces. The pipe's wall thickness determines its pressure rating. The optimal pipe size should be determined by minimising the sum of energy and piping costs.

Pipe sizes are standardised to reduce the number of pipe sizes. In current practice, pipe size is defined with two sets of numbers. They are: 1) Pipe schedule (wall thickness) and 2) Nominal pipe size (NPS) or Nominal diameter (DN). NPS is also referred to as NB (Nominal Bore).

Pipe Wall Thickness – Schedule Numbers

Manufacturers offer standard and non-standard sizes. Schedule numbers 40, 80, and 160 are most commonly used to specify the wall thickness of pipes in hydraulic systems.
- The schedule number 40 conforms to the 'standard' wall thickness intended for low pressures.
- The schedule number 80 conforms to the 'extra heavy' wall thickness intended for high pressures.
- The schedule number 160 conforms to the 'double extra heavy' wall thickness.

The schedule number indicates the approximate value of the expression $1000 \times P/S$, where P is the service pressure, and S is the allowable stress, both stated in the same unit.

Nominal Pipe Size (NPS)

It is a size standard established by the American National Standards Institute (ANSI). It is the number that defines the pipe's size. For a 6 NPS pipe, the 6" is the nominal size. Nominal sizes of pipes do not always correspond to its inside diameter or outside diameter. Hydraulic pipes come in nominal sizes from 1/8" to 42". Each size is available in a variety of wall thicknesses. Table 2.1 presents the minimum wall thicknesses for steel pipes of different sizes.

Table 2.1 | Standard nominal pipe sizes and dimensions

Nominal Pipe Size		Outside Diameter	Wall thickness		
			Schedule 40	Schedule 80	Schedule 160
--	inch	inch	inch	inch	inch
⅛	0.125	0.405	0.068	0.095	--
¼	0.250	0.540	0.088	0.119	--
⅜	0.375	0.675	0.091	0.126	--
½	0.500	0.840	0.109	0.147	0.188
¾	0.750	1.050	0.113	0.154	0.219
1	1.000	1.315	0.133	0.179	0.250
1¼	1.250	1.660	0.140	0.191	0.250
1½	1.500	1.900	0.145	0.200	0.281
2	2.000	2.375	0.154	0.218	0.344
2½	2.500	2.875	0.203	0.276	0.375
3	3.000	3.500	0.216	0.300	0.438
3½	3.500	4.000	0.226	0.318	--
4	4.000	4.500	0.237	0.337	0.531
5	5.000	5.563	0.258	0.375	0.625
6	6.000	6.625	0.280	0.432	0.719
8	8.000	8.625	0.322	0.500	0.906
10	10.00	10.75	0.365	0.594	1.125
12	12.00	12.75	0.406	0.688	1.312
14	14.00	14.00	0.438	0.750	1.406
16	16.00	16.00	0.500	0.844	1.594
18	18.00	18.00	0.562	0.938	1.781

As shown in Table 2.1, for pipes with nominal sizes ranging from 1/8" to 12", the internal diameter (ID) differs from the nominal size and varies with the pipe schedule number. For a pipe with a nominal pipe size of 14" or above, the pipe OD corresponds to the nominal pipe size. It may be noted that any increase in the wall thickness decreases the inside diameter of the pipe.

What is Nominal?
The term nominal indicates that the given dimension is as expected or approximate. It refers to a standard name that differs from the actual measured size. As we know, the internal diameter of a pipe is critical because the fluid flows through its interior. However, standards specify the outside dimensions to ensure the pipe is manufactured in standard sizes for easy fitting. The internal dimensions vary with pipe wall thickness. Any measurement of internal diameter can only be nominal.

Pipe Schedule for Stainless Steel Pipe
The cost of stainless steel pipe is much higher than that of carbon steel pipe. Due to stainless steel's corrosion-resistant properties, the development of high-alloy stainless steels and the use of fusion welding enable thinner steel pipes to perform satisfactorily without fear of early failure.

To reduce the cost of stainless steel pipe and fittings, the American Society of Mechanical Engineers (ASME) has introduced various schedule numbers. Under ASME B36.19, a schedule number with an 'S' suffix is introduced for stainless steel pipes. Schedule numbers 5S, 10S, 40S, 80S are used for the wall thickness of stainless steel pipe as per ASME B36.19.

Example 2.1
Refer to Table 2.1. Calculate the IDs of schedule 40 pipes corresponding to the following nominal pipe sizes: (1) 2" and (2) 14."

Solution
1.

NPS	= 2"
OD	= 2.375"
Wall thickness (t)	= 0.154
ID	= OD – 2 x t
	= 2.375 – 2 x 0.154
	= 2.067"

2.

NPS	= 14"
OD	= 14"
Wall thickness (t)	= 0.438
ID	= OD – 2 x t
	= 14 – 2 x 0.438
	= 13.124"

Diameter Nominal (DN)

It is the metric equivalent of NPS used to specify pipe sizes. It is based on millimeters (mm). Note that the metric designations conform to ISO and European standards. These standards ensure consistency and compatibility across manufacturers.

NPS is primarily used in the United States, where it is the standard for specifying pipe sizes across various industries.

DN is widely used in many countries outside the United States as the standard pipe size system.

Table 2.2 gives the standard metric pipe sizes and dimensions (Nominal diameters and wall thicknesses).

Table 2.2 | Pipe sizes in Diameter Nominal and wall thicknesses

Nominal Diameter (DN)	Outside Dia, mm	Wall thickness, mm				
		A	B	C	D	E
6	10.2	1.6				
8	13.5	1.8				
10	17.2	1.8				
15	21.3	2.0	2.8			
20	26.9	2.0	2.8			
25	33.7	2.0	3.2	4.2	6.3	6.3
32	42.4	2.3	3.5	4.2	6.3	6.3
40	48.3	2.3	3.5	4.2	6.3	6.3
50	60.3	2.3	3.8	4.2	6.3	6.3
65	76.1	2.6	4.2	4.2	6.3	7.0
80	88.9	2.9	4.2	4.2	7.1	7.6
90	101.6	2.9	4.5	4.5	7.1	8.1
100	114.3	3.2	4.5	4.5	8.0	8.6
125	139.7	3.6	4.5	4.5	8.0	9.5
150	168.3	4.0	4.5	4.5	8.8	11.0
200	219.1	4.5	5.8	5.8	8.8	12.5
450	457.0	6.3	6.3	6.3	8.8	12.5

Pipe Standards
1. Steel pipes according to DIN Standards
- DIN EN 10305-4. E 355N (St. 52.4 NBK) or E 235N (St. 37.4 NBK)
- DIN EN 10216-5 Seamless steel tubes
- Seamless cold-drawn precision steel pipes E235 (St 37.4) and E355N(St 52.4) according to EN 10305-4
- E235N (St 37.4) – NBK – Normalized, phosphated, and oiled inside and outside
- E235N (St 37.4) – NBK/ZN – Normalized, electric zinc plated with Cr-VI-free passivation
- E355N (St 52.4) – NBK – Normalized, phosphated, and oiled inside and outside
- E355N (St 52.4) – NBK/ZN – Normalized, electric zinc plated with Cr-VI-free passivation

2. Austenitic stainless steel pipes, seamless, cold-drawn 316L (1.4404) according to ASTM-standards
- AISI 316L – metric sizes fully annealed, scale-free
- TP 316L – schedule sizes fully annealed, scale-free

Pipe Fittings
Pipe connections are coupled using welded, flanged, or threaded joints. The jointing technology is selected based on working pressure, pipe size, pipe material, fitting standards, and other factors, such as potential pressure surges in the system and environmental conditions. Sections of pipes can be joined together using pipe fittings, such as sleeves, elbows, tees, bends, etc. (Figure 2.1). Welded connections are more commonly used in systems that involve severe mechanical loads, high pressure, vibration, or high temperature. They can also be employed where leaks cannot be tolerated. Fittings are available to various standards, including NPTF and JIC.

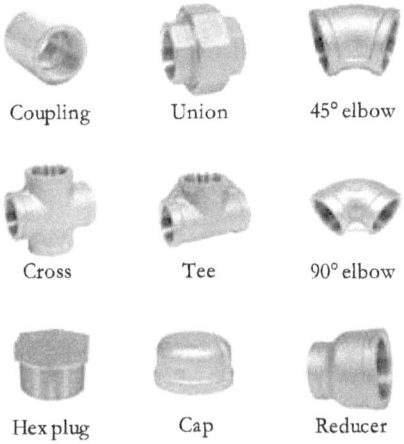

Coupling	Union	45° elbow
Cross	Tee	90° elbow
Hex plug	Cap	Reducer

Figure 2.1 | An assortment of pipe fittings

The welded connections can be butt-welded, socket-welded, or slip-on sleeve-welded. The socket weld is used for pipes less than 2 inches in diameter, whereas the butt weld is used for larger pipes.

However, in hydraulic piping systems with high-quality requirements, it is recommended to use non-welded connection technologies (fittings, flanges, etc.) for all pipe sizes due to their reliability and inherent cleanliness.

Threaded connections are the most common and are used in applications at pressures up to 170 bar.

Welded Pipe Joints
Butt-welded joints with complete root penetration can be used to join pipe sections. Socket-welded joints or slip-on welded sleeve joints, as shown in Figure 2.2(a), may be used for piping with a nominal diameter up to 80 mm, except for toxic or corrosive services. For slip-on welded sleeve joints,

the pipe sections are inserted into the sleeve and welded, as shown in Figure 2.2(b). The fillet weld leg size must be at least 1.1 times the pipe's nominal wall thickness.

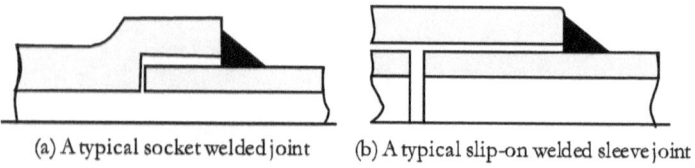

(a) A typical socket welded joint   (b) A typical slip-on welded sleeve joint

Figure 2.2 | Welded pipe joints

Flanged Joints

Pipe sections can also be joined with flanges, especially for pipes over 7/8" OD. Flanged joints consist of two mating flanged components sealed with an O-ring and bolted together. Flanges are attached to the pipes by welding (Figure 2.3) or using tapered threads.

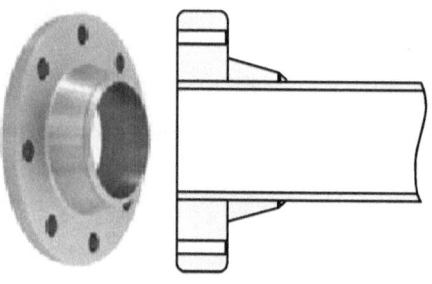

Figure 2.3 | Flanged joints

Flanged joints are robust and leak-free. They also offer easy maintenance and high vibration resistance. They are used in high-pressure hydraulic systems, particularly for joining large-diameter pipes or connecting pipes to high-torque-inducing components such as valves and pumps.

Thread Joints for Pipes

Thread joints are used for hydraulic service to produce a leak-proof metal-to-metal seal. They are either tapered or straight. The pipe threads are made pressure-tight by sealing them.

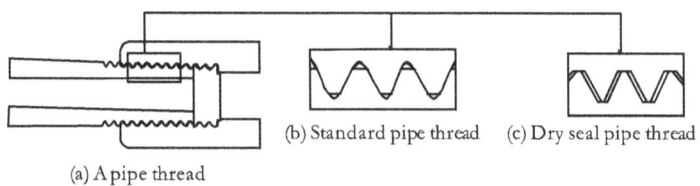

(b) Standard pipe thread   (c) Dry seal pipe thread

(a) A pipe thread

Figure 2.4 | Pipe threads

Pipe threads used in hydraulic piping can be divided into two types: (1) Standard pipe threads and (2) Dry-seal pipe threads. Figure 2.4 shows these types of pipe threads.

Standard Pipe thread

The standard pipe thread has tapered threads that produce a metal-to-metal seal, as shown in Figure 2.4(b). The taper is 1/16 of an inch. The connection is pressure-tight due to the sealing on the threads. This type of thread leaves a spiral clearance as the pipes are tightened. Pipe threads require sealant, such as Teflon tape or joint compound, to fill any voids between the threads and make the joint leakproof. However, these threads often develop leaks gradually, which are difficult to repair in the field.

Tapered pipe threads are used for standard hydraulic services (excluding toxic or corrosive fluids) at pressures up to 103 bar and temperatures up to 495°C. They can be used for hydraulic services, including connections to equipment such as pumps, valves, cylinders, accumulators, gauges, and hoses.

Dry-seal pipe thread

When larger pipes are used, dry-seal pipe threads are most appropriate. In this type of thread, pressure-tight joints are not made on the threads. Both threads are parallel, and sealing is achieved by compressing a soft material onto the external thread, as shown in Figure 2.4(c). When tightened, the dry-seal thread eliminates the spiral clearance. This type of thread form tends to minimise thread leaks.

Straight-thread with 'O' Ring

Straight-thread O-ring-type fittings, as shown in Figure 2.5, may be used for equipment connections without pressure or service limitations. This type may not be used to join pipe sections. British Standard Pipe Parallel (BSPP) and SAE Straight Thread O-Ring are included in this category.

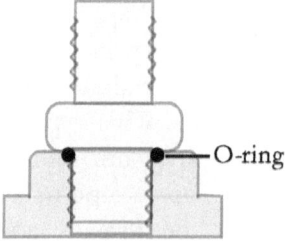

Figure 2.5 | Straight-thread with 'O' ring

Straight-thread O-ring-type fittings are recommended for medium- to high-pressure systems. The O-ring provides a superior, reliable seal. Major advantages of this type of connection include leak prevention, reusability, and ease of installation.

Basics of Threads

Hydraulic pipe threads, as shown in Figure 2.6, are specialized threaded connections that create leak-free, high-pressure seals in hydraulic systems. A screw has a male thread (external), while the matching hole has a female thread (internal). Pipe threads are classified as either tapered or parallel (straight).

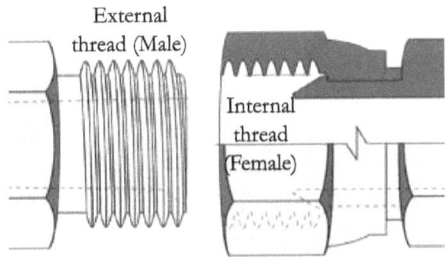

Figure 2.6 | Male and female threads

Parallel Threads and Tapered Threads

Figure 2.7 shows parallel and tapered threads. Parallel threads maintain a constant diameter along their length. Sealing in parallel threads is achieved via O-rings.

Tapered threads gradually decrease in diameter as they extend outward. Sealing in tapered threads is achieved by deforming the threads.

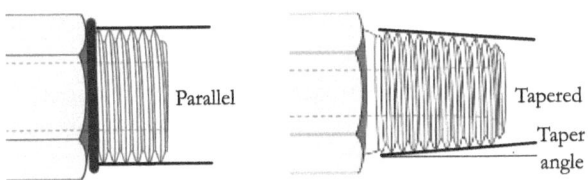

Figure 2.7 | Parallel thread and tapered thread

Terms and Definitions, Threads

Threads are governed by terms such as crest, root, threads per inch (TPI), major diameter, minor diameter, pitch diameter, flank angle, and taper angle. These terms are briefly explained below and illustrated in Figure 2.8.

Crest: The outermost part of a thread is called the crest.

Root: The innermost part of a thread is called the root.

Pitch: The pitch is the distance from the crest of one thread to the next, measured in mm.

Threads per Inch (TPI): Threads per inch is the number of thread peaks per inch of the screw.

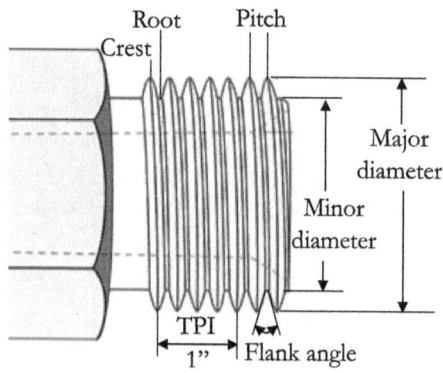

Figure 2.8 | Pipe thread

Major Diameter: The major diameter is defined by the thread tips.

Minor Diameter: The minor diameter is determined by the thread groove

Flank Angle: The flank angle is the angle between the flank of a screw thread and the perpendicular to the screw's axis.

Taper Angle: Tapered threads have a taper angle. This is the angle between the taper and the pipe's center axis.

Hydraulic Pipe Thread Types

Hydraulic pipe thread types include ISO coarse and fine threads (metric), British Standard Pipe (BSP) parallel and tapered, and National Pipe Thread (NPT). A type is chosen based on region and pressure requirements. These types and ISO metric screw threads are described below.

M - ISO Coarse Screw Threads (Metric)

The ISO metric screw thread is a globally standardized thread for fasteners and ports, as specified in ISO 724. An ISO metric thread can be coarse or fine. Table 2.3 presents the size chart for ISO metric coarse threads (M-series).

Table 2.3 | Size chart for M- ISO coarse screw thread

Thread size (mm)	Major diameter (mm)	Minor Diameter (mm)	Pitch (mm)
M3	2.98	2.459	0.5
M4	3.978	3.242	0.7
M5	4.976	4.134	0.8
M6	5.974	4.917	1.0
M8	7.974	6.917	1.0
M10	9.968	8.376	1.5
M12	11.97	10.106	1.75
M16	15.96	13.835	2.0
M20	19.96	17.294	2.5
M24	23.95	20.752	3.0

M - ISO Fine Screw Threads (Metric)

An ISO fine-pipe thread is a standardised metric thread with a smaller pitch than a coarse thread, used for precise connections and often designated as M-series. Table 2.4 presents the size chart for ISO Metric fine threads.

Table 2.4 | Size chart for M- ISO fine screw thread

Thread size (mm)	Major diameter (mm)	Minor Diameter (mm)	Pitch (mm)
M3x0.35	2.981	2.621	0.35
M4x0.5	3.978	3.242	0.5
M5x0.5	4.98	4.459	0.5
M6x0.75	5.978	5.188	0.75
M8x0.75	7.978	7.188	0.75
M10x0.75	9.978	9.188	0.75
M10x1	9.974	8.917	1.0
M10x1.25	9.972	8.647	1.25
M12x1	11.97	10.917	1.0
M12x1.5	11.97	10.376	1.5
M16x1	15.97	14.917	1.0
M16x1.5	15.97	14.376	1.5
M20x1	19.97	18.917	1.0
M20x1.5	19.97	18.376	1.5
M20x2	19.97	17.835	2.0
M24x1.5	23.97	22.376	1.5

ISO Metric (M-Series) Threads Notation

The metric designation of a screw thread is indicated by the letter M, followed by the nominal diameter in mm and then the pitch in mm—for example, M 20 x 1.5. For coarse thread, the pitch may or may not be displayed, whereas for fine thread, the pitch must be displayed.

BSPP Threads

The common types of British Standard Pipe (Whitworth) threads are BSPP(G)-Parallel and BSPT (R/Rp)-internally tapered/parallel. An appropriate sealing compound can be applied to the thread to ensure a leak-proof joint. Table 2.5 presents the BSPP (G) size chart (ISO 228).

Table 2.5 | Size chart for BSPP (G) threads

Thread Size (inch)	Major diameter (mm)	Minor Diameter (mm)	Threads Per Inch (TPI)
G 1/16	7.723	6.561	28
G 1/8	9.728	8.566	28
G 1/4	13.157	11.445	19
G 3/8	16.662	14.950	19
G 1/2	20.955	18.631	14
G 3/4	26.441	24.117	14
G 1	33.249	30.291	11
G 2	59.614	56.656	11

BSPT Threads

Table 2.6 presents the BSPT thread-size chart (ISO 7).

Table 2.6 | Size chart for BSPT threads

Male Thread Size (inch)	Female Thread Size (inch)	Major diameter (mm)	Minor Female Diameter (mm)	Threads Per Inch (TPI)
R 1/16	Rp 1/16	7.723	6.490	28
R 1/8	Rp 1/8	9.728	8.495	28
R 1/4	Rp 1/4	13.157	11.341	19
R 3/8	Rp 3/8	16.662	14.846	19
R 1/2	Rp 1/2	20.955	18.489	14
R 3/4	Rp 3/4	26.441	23.975	14
R 1	Rp 1	33.249	30.111	11

NPT Pipe Threads

National Pipe thread is the USA standard for tapered threads. The most common types of National Pipe Threads are National Pipe Taper (NPT) and National Pipe Taper Fuel (NPTF). NPTF is the taper pipe thread for a dry-seal joint without sealant compound. The NPT pipe thread size chart is given in Table 2.7.

Table 2.7 | NPT threads size chart

Thread Size	Major Diameter (mm)	Threads Per Inch (TPI)
1/16" – 27 NPT	7.938	27
1/8" – 27 NPT	10.287	27
1/4" – 18 NPT	13.716	18
3/8" – 18 NPT	17.145	18
1/2" – 14 NPT	21.336	14
3/4" – 14 NPT	26.670	14

Flank Angle

The flank angles for ISO, BSPP, BSPT, and NPT are shown in Figure 2.9.

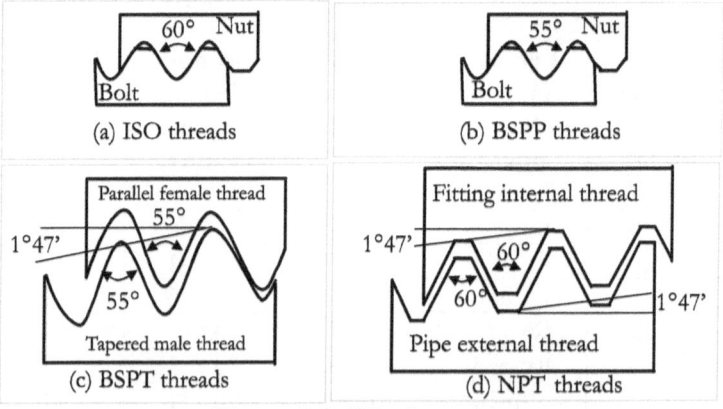

Figure 2.9 | Flank angles

Standards

Important standards for various thread types are given in Table 2.8.

Table 2.8 | Standards pertinent to threads.

Threads	Designation	Type	Standard
M - ISO screw thread (Metric)	M	Coarse thread	ISO 724 (DIN 13-1)
	M	Fine thread	ISO 724 (DIN 13-2 to 11)
NPT – Pipe thread	NPT		ANSI B1.20.1
	NPTF		ANSI B1.20.3
G/R/R_P – Whitworth pipe thread	G	BSPP	ISO 228 (DIN 259)
	R / Rp	BSPT	ISO 7 (EN10226)
UNC/UNF – Unified National Thread	UNC		ANSI B1.1

Comparison

A comparison of various thread types by type, flank angle, thread angle, and seal location is presented in Table 2.9.

Table 2.9 | Comparison of threads.

Thread type Parameter	Metric (Standard & Fine)	BSPP (G)	BSPP (R)	NPT	UNC / UNF
Type	Parallel	Parallel	Tapered	Tapered	Parallel
Flank angle	60°	55°	55°	60°	60°
Thread angle	0°	0°	1°47'	1°47'	0°
Seal location	-O-ring -Gasket	-O-ring -Gasket	On threads	On threads	-O-ring -Gasket

Pipe Supports

When designing supports for piping, the following factors should be taken into account:

- A pipe shall not be supported by other pipes, nor should the pipes be utilised to support other components.
- The transfer of vibration from other equipment and machinery should be avoided to the extent possible.
- Thermal expansions shall be taken into account when designing the supports.
- A bend should be supported as close to the bend as possible (whenever possible on both sides of the bend).
- The support should be located as close to the end of the pipe as possible when connecting to the hose.

Tests

All piping systems are to be tested under working conditions after installation and a visual inspection for cracks, corrosion, and physical damage. Common testing methods include hydrostatic, impulse, and destructive testing.

Hydrostatic Testing: After fabrication, pipes and integral fittings may be hydrostatically tested at 1.5 times the design pressure to detect leaks and pipe deformation. Small pipes with an outside diameter of less than 15 mm are usually exempt from hydrostatic testing.

In an **impulse test**, the pipe is subjected to repeated pressure pulses to evaluate its fatigue life and durability.

In a **destructive test**, pipe samples or hoses are subjected to very high pressure until they burst. A test sample may also be subjected to normal pressure for an extended period to assess its long-term strength.

Chapter 3 | Hydraulic Tubes

A tube is the most widely used type of hydraulic system conductor. The tube is generally a small-diameter, thin-walled pipe. It can be bent into almost any shape, reducing the number of tube fittings required when configuring a conductor system. Note, a conductor system with a tube is easier to handle. However, it is usually more expensive than the piping system due to its tighter manufacturing tolerances.

Tube Construction
A seamless tube, as shown in Figure 3.1, is formed by cold drawing a billet over a piercing rod. A welded tube is made by forming a piece of cold-rolled steel into a tube and then joining the longitudinal seam using a material-fusion process. The tube may have to be pre-formed or bent before installation. Although the tube may provide a neater appearance when installed correctly, bending it requires careful planning and skill.

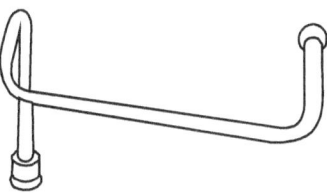

Figure 3.1 | A tube section

Specifications of Tubes
Essential specifications for hydraulic tubes include their size, pressure ratings, and minimum bend radius. Tube size is always specified by its outside diameter (OD). Available tube sizes in metric units range from 4 mm to 42 mm.

Typical Size and Pressure Chart for Seamless Cold-drawn Steel Tubes

Table 3.1 | Pressure chart for seamless cold-drawn tubes

Tube OD (mm)	Wall thickness (mm)	Max. Working pressure (bar)	Theoretical burst pressure (bar)
6	1.0	389	1680
8	1.0	333	1190
8	1.5	431	1860
10	1.0	282	870
10	1.5	373	1380
12	1.5	353	1150
12	2.0	409	1580
14	2.0	403	1340
15	1.5	282	980
16	1.5	264	820
16	2.0	353	1170
18	1.5	235	780
18	2.0	303	948
20	2.0	222	920
20	2.5	353	1220
22	2.0	256	850
25	2.0	226	670
25	2.5	282	920
25	3.0	338	1050
28	2.0	201	620
30	3.0	284	920
30	4.0	376	1250
35	2.5	201	620
38	4.0	297	970
38	5.0	371	1350
42	3.0	201	1580

Pressure Rating, Tubes

Burst pressure is the pressure at which a tube ruptures. Working pressure is the value considered safe for operating the system under normal working conditions. The safety factor is the ratio of the working pressure to the burst pressure. As a general rule, a good safety factor is 4:1 or greater.

Tube Material

Tube is constructed of dead-soft, cold-drawn carbon steel (St 37.4 and St 52.4) and has become the accepted standard for hydraulics because it has the mechanical strength required to withstand high pressures.

St 37.4 is a low-carbon, cost-effective, seamless tube, often finished as NBK (Normalised, Black, oiled), with a tensile strength of 3400 to 4800 bar.

St 52.4 steel is a high-strength, high-quality, low-alloy, non-alloy carbon steel with a tensile strength in the range of 4900 to 6300 bar. If greater strength is required, the tube can also be made from AISI 4130 steel, which has a tensile strength of >5000 bar.

Example 3.1

Determine the inside diameter of a tube section with an OD of 20 mm and a wall thickness of 2.5 mm.

Solution

Pipe OD	= 20 mm
Wall thickness	= 2.5 mm
Inside diameter of the pipe	= $D_o - 2t$
	= 20 − 2 x 2.5
	= 15 mm

Example 3.2

Determine the hoop stress in a tube with an outside diameter of 22 mm and a wall thickness of 2 mm when subjected to a pressure of 200 bar.

Solution

Outside diameter of the tube, D_i = 22 mm
Wall thickness of the tube, t = 2 mm
Pressure, P = 70 bar

Hoop stress developed = $P \times D_i / 2t$
= 70 x 22 / 2 x 2 = 385 bar

Example 3.3

What would be the maximum pressure rating of a tube of OD of 30 mm with a wall thickness of 3 mm? The tube material is stainless steel with a tensile strength of 5500 bar. Assume a safety factor (SF) of 4.

Solution

Tube OD, Do = 30 mm
Wall thickness, t = 3 mm
Tensile strength, S = 5500 bar
Safety factor, SF = 4

Tube ID, Di = Do – 2 x t
= 30 – 2 x 3 = 24 mm

Burst factor (BF) = 2ts / Di
= 2 x 3 x 5500 / 24
= 1375 bar

Tube pressure rating = BP / SF
= 1375 / 4 = 344 bar

Minimum Bend Radius

It is the smallest radius of the curved section of a fluid conductor (tube or hose) beyond which it cannot be bent without flattening, kinking, or wrinkling. Bending the conductor beyond this limit causes excessive flattening of the tube in the bend region and internal damage, leading to premature failure. It also causes severe backpressure. The bend radius is typically measured along the tube's centerline, as shown in Figure 3.2. It is measured as the distance from the tube's center of curvature to its centerline. A rule of thumb suggests a minimum bend radius of three times the outside diameter.

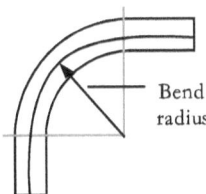

Figure 3.2 | Minimum bend radius of tubes

Tube Bending Process

A hydraulic tube can be bent by hand or with power bending equipment. Steel tubes can be bent using methods such as roll forming, press forming, mandrel bending, or table forming. In a tube-bending process, pressure is applied to bend the tube around a correctly sized die to achieve the required radius.

Advantages and Disadvantages, Tubes

The main advantage of tubes is that they can be bent into shape, reducing the number of fittings. Fewer connections generally mean a lower risk of leaks. Tube is also known for its ability to absorb vibration and for its smooth interior finish, which is suitable for fluid flow.

Example 3.4

Select the appropriate metric-size steel tube for a flow rate of 0.0019 m³/s and an operating pressure of 70 bar. The maximum recommended velocity is 6 m/s, and the tube material is SAE 1010 dead-soft, cold-drawn steel with a tensile strength of 380 MPa. Assume a safety factor (SF) of 6.

Solution

Flow rate, Q   = 0.0019 m³/s
Operating pressure, P   = 70 bar
Velocity, v   = 6 m/s
Tensile strength, S   = 380 MPa = 3800 bar

Tube cross-sectional area, A = Q/v
                             = 0.0019 / 6
                             = 0.0003167 m²

Tube ID, d (at least)   = $\sqrt{(4 \times A / \Pi)}$ = 20 mm

Selected tube OD   = 25 mm
Wall thickness   = 2 mm
Tube ID   = OD – 2xID = 25 – 2x2 mm
                        = 21 mm

Burst Pressure (BP)   = 2ts/Di
                        = (2 x 0.002m x 3800 bar)/0.021m
                        = 724 bar

Working Pressure   = BP/SF = 724/6 = 120 bar

The working pressure exceeds the operating pressure; therefore, this tube (25 mm OD, 2 mm wall thickness) is suitable for use.

Tube Fittings

Because a tube's wall thickness is relatively thin, threading cannot be used to seal the tube connections. A variety of tube fittings are available for hydraulic applications. Tubes can be joined quickly and easily with flaring, brazing, or couplings. Flared or flareless fittings are used for tube end connections.

For tube ODs from 10 to 22 mm and pressures up to 210 bar, flareless end-forming connection technology is used. For tube ODs from 28 to 42 mm, 37° or 45° flaring is generally used. A tube is usually flared to JIC 37° (for >70 bar) or SAE 45° (for <70 bar). The 45° flare angle is used in automotive and refrigeration work, but not in hydraulic plumbing.

Flare Fitting, Tube

It consists of a nut, a sleeve over the flared tube, and a body, as shown in Figure 3.3. When forming flares, it is necessary to prepare the tube: cut it square, file it smoothly, and remove burrs.

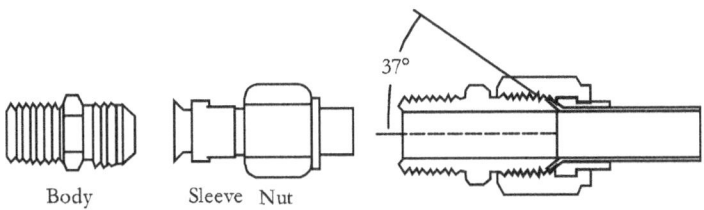

Figure 3.3 | Flare fitting

The most critical step in making a flare tube fitting is forming the flare without galling, over-thinning, or splitting the tube end. The sleeve and nut are pushed smoothly over the tube end. The sleeve prevents the nut from twisting during tightening. When the nut is screwed onto the body, it draws the sleeve and the flare against the body, thus forming a seal.

Compression (flareless) Fitting, Tube
It consists of a body, ferrule(s), and a nut, as shown in Figure 3.4. First, the ferrules and nut should be slipped over the tube. The tube is inserted into the body, where it butts up against the shoulder. When the nut is screwed onto the body, the ferrule bites into the tube's skin to provide holding strength for the connection. This tight connection provides a positive seal. It is used on medium- and heavy-wall tubes or when the tube cannot be flared.

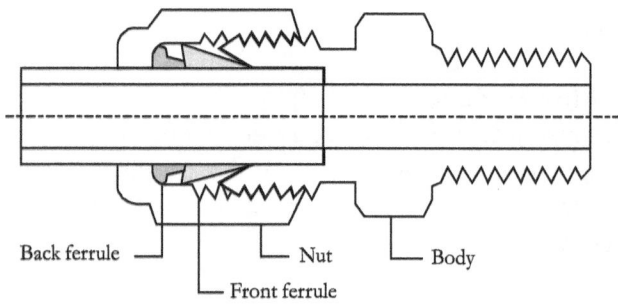

Figure 3.4 | Compression fitting

Selection of Tubes and Fittings
Proper tube material, type, and size for a given application and fitting type are critical for the efficient, trouble-free operation of the associated fluid power system. Selecting the proper tube and fittings involves choosing the appropriate tube material and determining the optimal tube size (O.D. and wall thickness). Proper tube sizing across various parts of a hydraulic system results in an optimal balance of efficiency and cost-effectiveness.

Chapter 4 | Hydraulic Hoses

Hoses are the most flexible and versatile type of conductors. They bend and flex easily. They can withstand vibration and pulsation better than tubes. They are selected when rigid pipes or semi-rigid tubes cannot be used, such as in applications involving components that move relative to each other or experience excessive vibration and constant pressure pulsations.

A hose assembly consists of a hose and end fittings that connect directly to adjoining pipe-work or fittings. A hose assembly must have the correct end-fitting configurations. It is easy to install a hose assembly with a well-thought-out routing layout. Given its superior routing flexibility, a hose is generally preferred over a metal tube.

Hose Construction

Hydraulic hoses consist of three parts: an inner tube, a reinforcement layer(s), and a protective cover, as shown in Figure 4.1.

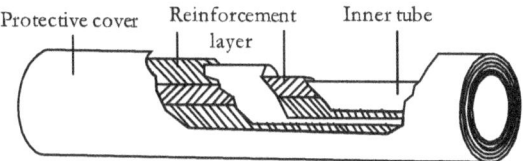

Figure 4.1 | A cut section of a hose

Inner Tube

The inner tube is the hose lining, which comes into direct contact with the fluid. Therefore, it must be chemically resistant to the fluid. It must also withstand extreme variations in fluid temperature. It is made of synthetic rubber, thermoplastics, or PTFE.

Reinforcement Layer(s)

The reinforcement section, as shown in Figure 4.2, provides the strength to withstand internal pressures and external forces. This section is constructed with one or more layers of braids or spirals. A reinforcement layer can be made of steel wire, textile, synthetic material, or a combination of these.

Figure 4.2 | Different types of hose reinforcement layers

The type of reinforcement depends on the hose's intended use. A hose can be made with up to six layers to meet the most demanding applications. Spiral reinforcement is particularly well-suited to high-pressure impulse applications.

The reinforcement layer of a hose connected to the suction side of a pump can also be made with a helical coil to prevent the hose from collapsing under the partial vacuum during fluid suction.

A hose with multiple reinforcements may be provided with an anti-friction layer between them to prevent the steel wires from rubbing against each other.

Protective Layer

The primary purpose of the cover is to protect the tube and reinforcement from abrasion, corrosion, extreme temperatures, UV light, and ozone. The cover can be made from synthetic rubber, fibre braids, or a combination of both, depending on the application. Hoses with synthetic rubber covers are preferred over those with textile-braid covers because they are more abrasion-resistant.

Summary of Hose Layers

A summary of the properties and materials of hose layers is presented in Table 4.1.

Table 4.1 | Properties and materials of hose layers

Hose layer	Properties	Material
Inner tube	-Direct contact with fluid -Must be chemically resistant to fluid -Must withstand temperature variations	-Synthetic rubber -Thermoplastics -PTFE
Reinforcement layer(s)	-Determines the working pressure -A layer is braided, spiraled, or coiled -Constructed with a single layer or multiple layers	-Steel wire -Textile -Synthetic -Combination
Protective layer	-Protects the tube and reinforcement from abrasion, corrosion, extreme temperatures	-Synthetic rubber -Fibre braids -Combination

Types of Hoses by Operating Pressures

Another way to classify hydraulic hoses is by pressure rating. The classification is presented below:

Low-pressure hoses are designed for use in various applications with operating pressures below 20 bar. Their reinforcement is usually textile.

Medium-pressure hoses: They are used for hydraulic applications requiring operating pressures of 20 to 210 bar. They may be constructed from a single wire braid or multiple wires and/or a textile braid.

High-pressure hoses: They are commonly used in hydraulic systems for construction equipment that operate at 210-415 bar. These hoses are often called 'two-wire' braid hoses because they generally have a reinforcement of two-wire braids of high-tensile-strength steel.

Very High-pressure Hoses: They are used for off-highway equipment and heavy-duty machinery that experience extremely high-pressure surges. The oil-resistant synthetic tubes in these hoses are reinforced with four or six layers of spiralled, high-tensile steel wire over a layer of yarn braid.

Suction Hose Pressure: A hose connected to a pump suction service is subjected to crushing forces because the atmospheric pressure outside the hose is higher than the internal pressure. Therefore, a hose connected to the suction line must withstand the pressure differential across it. The best way to prevent the hose from collapsing is to reinforce the hose with a helical wire.

Hose Size Specifications

Choose a hose with an internal diameter sufficient to minimise pressure loss and prevent hose damage from excessive fluid turbulence. A small hose increases pressure loss. An oversized hose adds cost, size, and weight.

The essential specifications of a hose include the inside diameter (ID), wall thickness (t), pressure ratings, and minimum bend radius.

Inside Diameter, Hose

Hose size is specified by its inside diameter (ID). The ID must be sufficient to provide the proper fluid volume within the permissible pressure drop for the specific application. The ID is specified in dash sizes or metric units.

As mentioned in a previous section, a dash number indicates the hose size in 1/16-inch increments. For example, a hose with a 1/4" (=4/16") inside diameter indicates that it has four numbers of 1/16" segments. Therefore, its dash size would be -4 (dash 4).

Pressure Rating, Hose

The pressure rating of a hose is determined by its construction. The pressure rating is determined by the number of layers, the materials, and the construction method. The more reinforcement layers a hose has, the higher the pressure it can withstand.

Burst pressure is attained when the hose ruptures or when leakage occurs from the end fitting. The rated pressure of a hose must be higher than the normal system pressure, and any pressure surges it will encounter.

Minimum Bend Radius, Hoses

The minimum bend radius is an essential consideration in the design and selection of a hose. Table 4.1 gives the typical parameters of hoses.

Table 4.1 | Typical parameters of hoses in metric units

Dash Number	ID		Work pressure bar	Min. Burst pressure bar	Min. bend radius mm
	Inch	mm			
-2	1/8	3.2	210	1088	
-3	3/16	4.8	210	1088	
-4	1/4	6.4	210	1088	38.1
-5	5/16	7.9	210	1088	
-6	3/8	9.5	210	1088	63.5
-8	1/2	12.7	210	1088	73.7
-10	5/8	15.9	210	1088	83.8
-12	3/4	19.0	210	1088	101.6
-14	7/8	22.2	210	1088	
-16	1	25.4	210	1088	127.0
-20	1 1/4	31.8	210	1088	304.8
-24	1½	38.1	210	816	355.6
-32	2	50.8	210	816	
-36	2 1/4	57.6	210	816	
-40	2½	63.5	210	816	
-48	3	76.2	210	816	
-56	3½	88.9	210	816	
-64	4	101.6	210	816	
-72	4½	115.2	210	816	

Types of Hose Motions
Typical hose motions are shown in Figure 4.3.

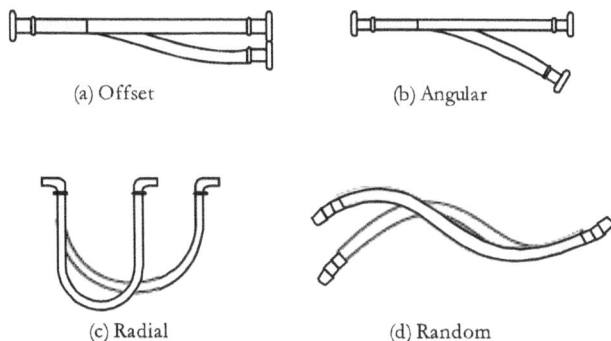

(a) Offset   (b) Angular

(c) Radial   (d) Random

Figure 4.3 | Types of hose motion

- When one end of the hose is moved in a plane perpendicular to its longitudinal axis, with its ends remaining parallel, it is called offset motion.
- The angular motion of the hose occurs when one end of the hose is moved in a simple bend, with its ends not remaining parallel
- The radial motion of the hose occurs when the hose is bent in a circular arc
- The random motion of the hose occurs in random planes

Standards of Hoses
ISO and SAE specify standards for hoses that define dimensional and performance parameters.

- ISO 1436-1, wire braid reinforced
- ISO 4079-1, textile reinforced
- ISO 3949, thermoplastic textile reinforced
- ISO 3862-1, spiral wire reinforced

The SAE J517 (US) standard defines the 100R series (SAE 100 R1 to SAE 100 R18), which specifies the construction, dimensions, pressure, and temperature ratings for hoses. Extracts of SAE 100 R2 and SAE 100 R18 specifications are presented below:

SAE 100 R2 Specifications

The hose shall consist of an inner tube of oil-resistant synthetic rubber, steel wire reinforcement, and an oil and weather-resistant synthetic rubber cover. A ply or braid of suitable material may be used over the inner tube and/or over the wire reinforcement to anchor the synthetic rubber to the wire.

SAE 100 R18 Specifications

The hose shall consist of a thermoplastic inner tube resistant to hydraulic fluids, with suitable synthetic fibre reinforcement, and a hydraulic-fluid- and weather-resistant thermoplastic cover.

Table 4.2 lists the number and type of braids and the pressure rating of hoses for various SAE numbers under SAE standards.

Table 4.2 | Braids and pressure ratings

SAE Number	Braids	Pressure (psi)
100R1	1 Wire	500 – 2500
100R2	2 Wire	1200 – 3500
100R3	2 Cloth	375 – 1250
100R4	1 Wire Spiral	50 – 300
100R6	1 Cloth	300 – 600
100R12	4 wire	2500 – 5000

In Europe, the Committee for European Normalisation (CEN/EN) standard specifies flexible hoses.
- EN 853 – Wire braided hose
- EN 854 - Fabric braided hose
- EN 855 – Thermoplastic textile braid hose
- EN 856 – Spiral wire hose

Selection of Hoses

Selecting the right hose is the first step towards the safe operation and long service life of the associated hydraulic system. It all begins by choosing the right components, such as hoses, couplings, crimping equipment, and accessories. Hoses must meet the requirements of size, pressure, bend radius, and routing. It is also necessary to know the equipment type, working and impulse pressures, the fluid to be used, and the bend radius.

Applications of Hoses

Hoses are used when rigid or semi-rigid pipes or tubing cannot be used, such as in applications involving moving machine parts.

Advantages and Disadvantages of Hoses

Hoses offer many advantages. They are flexible and portable, and they absorb and dampen pressure surges and vibration. It is easier and faster to route hoses, even around obstacles, than with other types of conductors. They require no brazing or specialised bending.

However, mixing and matching couplings from one manufacturer with hoses from another manufacturer can lead to premature or catastrophic assembly failure. Hoses are susceptible to abuse, misapplication, and improper plumbing.

Hose Fittings

Fittings are metal components that are attached or crimped onto a hose to form a hose assembly, as shown in Figure 4.4. One end (hose end) of a fitting attaches to the hose, and the other end (thread end) connects the hose assembly to another component in a hydraulic system.

| End fitting | Hose | End fitting |

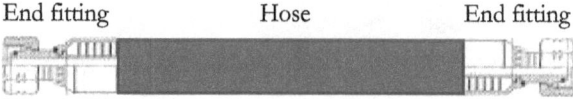

Figure 4.4 | Hose assembly

Hose fittings are made of carbon steel, stainless steel, or brass, depending on the application and cost. Carbon steel fittings can withstand pressures typically up to 350 bar. Stainless steel fittings can withstand pressures typically up to 700 bar.

Carbon steel fittings are more affordable but are not highly corrosion-resistant. Stainless steel fittings offer the best strength and durability. Brass fittings can withstand pressures up to 210 bar. Brass fittings have an excellent mix of strength and corrosion resistance. All these fittings have high-temperature resistance.

Hose-end Connections: A fitting's hose-end connection can be permanent or reusable.

Permanent Hose Fittings: They can be permanently secured to the hose by crimping a stainless steel swaged ferrule over the fitting's barbed hose tail, ensuring a reliable, resilient connection that is difficult to break. A crimping machine can be used to crimp the fitting.

Reusable Hose Fittings: They are field-installable fittings. Each consists of a female fitting (ferrule) with the same hose ID and a male adapter (stem), as shown in Figure 4.5.

The hose end of the ferrule is threaded onto the hose, and the male adapter is screwed into the female end of the ferrule. After assembly, the ferrule is pressed against the hose to form a secure seal.

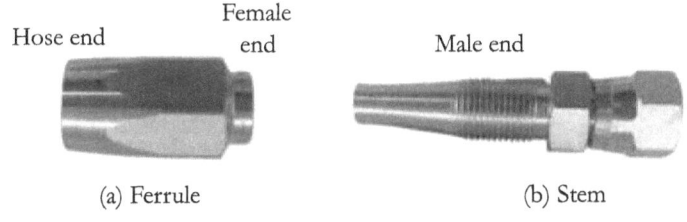

Hose end   Female end   Male end

(a) Ferrule   (b) Stem

Figure 4.5 | Reusable hose assembly

End-fittings: An end fitting is a mechanical connector that secures a hose to equipment or another hose, ensuring a leak-proof, secure connection. End fittings, as shown in Figure 4.6, include male- and female-threaded fittings and quick-release couplings tailored for specific applications.

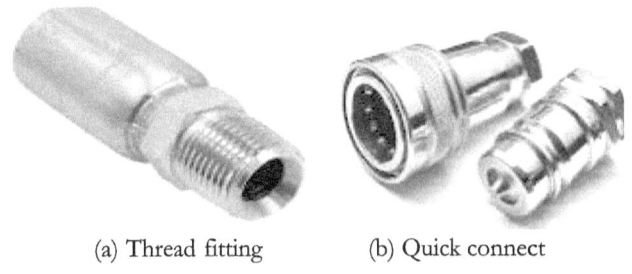

(a) Thread fitting   (b) Quick connect

Figure 4.6 | Hose end-fittings

Hose Fitting Standards

Fittings are made to metric or SAE standards. These standards provide general and dimensional specifications. BS EN ISO 12151-2 is the British standard, and J516 202111 is the SAE standard for hydraulic hose fittings.

Quick Couplings (or Disconnects)

They are used for convenience, as they can be installed and removed by hand, and in situations where repeated connection and disconnection of the lines is required. A quick coupling has a male side and a female coupler. Quick couplings can be poppet or flat-face types.

In the poppet type, the male poppet (nipple) gets depressed when it engages with the coupler. This action opens the valve, allowing hydraulic fluid to flow. Poppet-type quick couplers fall into ISO A or ISO B styles.

In flat-face couplers, both coupling sides are flat. A flat face male nipple will mate with a female flat face coupler. The back end of flat-face couplers can come with NPT, JIC, ORFS, or straight-thread O-ring threads.

Based on the coupling's valve configuration, hydraulic couplings generally fall into one of two groups: double-shutoff and straight-through.

Double Shutoff Couplings

They are used when it is essential to minimise fluid loss upon disconnection. Both halves of the coupler, the body and the nipple, contain shutoff valves, as shown in Figure 4.7. These valves open automatically when the body and nipple are connected and close automatically when the two halves are disconnected, minimizing fluid loss.

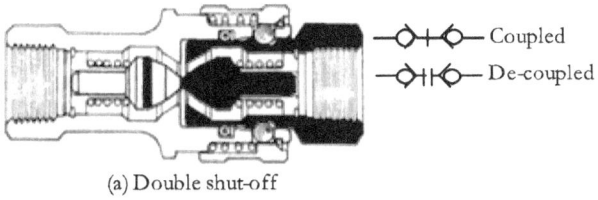

(a) Double shut-off

Figure 4.7 | A double shutoff coupling

Straight-Thru couplings

They have no valves in either half and are ideal for applications requiring maximum flow. Their smooth, open bore offers the lowest pressure drop of any quick-disconnect coupling and allows thorough cleaning.

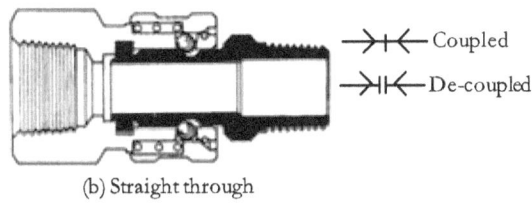

(b) Straight through

Figure 4.8 | A double straight-through coupling

Since there are no valves in either half, the fluid flow should be shut off before disconnecting the coupling.

Straight-through couplings are used where flow must be unrestricted. Figure 4.8 shows the cross-sectional view of a straight-through coupling.

Every manufacturer etches a part number on each coupler to aid identification. It is better to consult the manufacturer's reference guide to ensure that quick disconnects (QDs) are properly matched.

Chapter 5 | Design of Hydraulic Piping Systems

A piping system for a hydraulic application is to be designed in accordance with the application's many requirements. In general, the suction pipe should be short and straight, and the return line should be large enough to limit the backpressure.

When designing the piping system, the following factors have to be taken into account: (1) System pressure, (2) Operating temperature, (3) Duty cycle, (4) Shocks and vibration, (5) Material, (6) Connection technology, (7) Hoses and hose couplings, (8) Pipe supports, and (9) Standards.

Fluid Velocities – Suction Lines

The suction line is typically dimensioned so that the velocity does not exceed 1.2 m/s. The recommended fluid velocities for initial pipe sizing in suction lines are given in Table 5.1.

Table 5.1 | Fluid velocities in suction lines

Viscosity (mm^2/s)	Maximum velocity (m/s)
150	0.6
100	0.75
50	1.1
30	1.2

Fluid Velocities - Pressure Lines

The recommended fluid velocities for initial pipe sizing in pressure lines are given in Table 5.2.

Table 5.2 | Fluid velocities in pressure lines

Pressure line	For flow rate >10 lpm
63 – 100 bar	4.0 – 4.5
100 – 160 bar	4.5 – 5.0
160 – 250 bar	5.0 – 5.5
250 – 400 bar	5.5 – 6.0

Fluid Velocities – Return Lines
Fluid velocities for initial pipe sizing in return lines should be between 2 and 3 m/s.

Fluid Conductor Sizing
Hydraulic conductor sizing involves selecting the appropriate conductor ID to balance efficiency, pressure drop, and heat loss. An undersized conductor can cause turbulence and excess heat, while an oversized conductor can lead to sluggish performance. Conductor sizing can be based on flow velocity or permissible pressure drop. The following sections present these methods.

Dimensioning Based on Flow Velocity
When using a flow-velocity-based dimensioning method, a pipe inner diameter can be determined from the equation below, given the maximum flow rate and the recommended flow velocity.

$$d = \sqrt{\frac{4 \times Qmax}{\prod \times v}}$$

d – Inner diameter of the pipe (m)
Q_{max} – Maximum flow rate (m³/s)
v – Flow velocity (m/s)

Example 5.1
Determine the inner pipe diameter required to achieve a flow velocity of 4.5 m/s at a flow rate of 100 lpm.

Solution
Maximum Flow rate, Qmax = 100 lpm
$$= 100/60000 = 0.00167 \text{ m}^3/\text{s}$$

Velocity, v $= 4.5 \text{ m/s}$

$$\text{Pipe ID} = \sqrt{\frac{4 \text{ x Qmax}}{\Pi \text{ x v}}}$$

$$= \sqrt{\frac{4 \text{ x } 0.00167}{\Pi \text{ x } 4.5}}$$

$$= 0.0217 \text{ m} = 21.7 \text{ mm}$$

Dimensioning Based on Pressure Losses
When using the pressure-loss-based dimensioning method, the pipe inner diameter is selected to ensure the resulting pressure losses do not exceed a specified value. The total pressure loss is limited to 3-5% for systems operating continuously. The total pressure loss is limited to 7-10% for systems with an intermittent duty cycle.

Flow Types
When the flow is laminar, all fluid particles move parallel to the pipe. As flow velocity increases, the flow becomes turbulent, meaning the direction of individual fluid particles varies. The flow type can be determined by computing the Reynolds number (Re) and comparing it to the critical

Reynolds number. The Reynolds number can be determined with the equation:

$$Re = \frac{v \cdot d \cdot \varrho}{\mu} = \frac{v \cdot d}{v}$$

Where,
v = Flow velocity [m/s]
d = Inner diameter of the pipe [m]
v (nu) = Kinematic viscosity [m²/s]
μ = Absolute viscosity [Pa.s]

The flow is said to be laminar when Re < Re (critical). The flow may be treated as turbulent when Re > Re (critical). Re (critical) can be taken as 2000.

Example 5.2
Determine the inner pipe diameter required to achieve a flow velocity of 5 m/s at a flow rate of 100 lpm. A fluid with a density of 820 Kg/m³ and an absolute viscosity of 0.32 N.s/m² is flowing through the pipe. Is the flow laminar or turbulent?

Solution

Q	= 100 lpm
v	= 5 m/s
d	= 0.02122 m = 21.22 mm

$$Re = \frac{v \cdot d \cdot \varrho}{\mu}$$

$$Re = \frac{5 \times 0.02122 \times 820}{0.32}$$

Re = 272 (Laminar flow)

Example 5.3

Determine the inner pipe diameter required to achieve a flow velocity of 1 m/s at a flow rate of 100 lpm. A fluid with a density of 820 Kg/m³ and an absolute viscosity of 0.32 N.s/m² is flowing through the pipe. Is the flow laminar or turbulent?

Solution

$$V = 1 \text{ m/s}$$
$$\varrho = 820 \text{ Kg/m}^3$$
$$\mu = 0.32 \text{ N.s/m}^2$$
$$d = 0.046 \text{ m} = 46 \text{ mm}$$

$$Re = \frac{v \cdot d \cdot \rho}{\mu}$$

$$Re = \frac{1 \times 0.046 \times 820}{0.32}$$

$$Re = 117 \text{ (Laminar flow)}$$

Dimensioning Based on Pressure Losses

When a fluid flows through a pipe, friction between the pipe wall and the fluid causes a pressure loss. This pressure loss occurs in two forms, as the piping system consists of straight sections, bends, and junctions. One is the major pressure loss from friction in straight pipe sections, and the other is the individual pressure loss from fittings and valves. Both types can significantly impact the system's performance.

This pressure loss is an irreversible loss of the fluid's potential energy. Calculating this loss is fundamental to designing a pipeline system.

Frictional Pressure Losses

The pressure losses in pipes and hoses can be estimated from the Darcy-Weisbach equation as given below:

$$\Delta Pa = \lambda \cdot \frac{l}{d} \cdot \frac{\varrho \, v^2}{2}$$

Δp_a = Frictional pressure loss, [Pa]
λ = Frictional resistance factor [-]
l = Length of the pipe [m]
d = pipe id [m]
ϱ = Hydraulic fluid density [Kg/m^3]
v = Flow velocity [m/s]

Frictional Resistance Factor, λ

The friction factor is a function of Reynolds number and surface roughness for round pipes and can be determined either analytically or by consulting the classic Moody Diagram. For laminar flow, the friction factor is independent of the surface roughness, and for turbulent flow, the friction factor is dependent both on the Reynolds number and the surface roughness.

For Laminar Flow

If the flow is laminar, λ depends only on the Reynolds number. The friction factor is given by:

$$\lambda = \frac{64}{Re}$$

Frictional Pressure Losses in Straight Pipe Sections

The pressure loss in a pipe for laminar flows can be determined by using the following equations in terms of the average velocity of flow or the fluid flow rate (more favourable), respectively:

$$\Delta Pa = \frac{32 \, \mu \, l \, v}{d^2}$$

$$\Delta Pa = \frac{128 \, \mu \, l \, Q}{\pi \, d^4}$$

Note: The pressure drop scales linearly with line length; therefore, long or small-diameter lines could impact system efficiency.

Example 5.4
Calculate the frictional pressure loss for a 50 mm dia pipe of length 50 m through which a fluid is flowing at a velocity of 5 m/s. The kinematic viscosity of the fluid is 0.001 m²/s, and the density of the fluid is 820 Kg/m³.

Solution

v	= 5 m/s
l	= 50 m
D	= 50 mm
ϱ	= 820 Kg/m³
ν	=0.001 m²/s

Re $= vd/\nu = 5 \times 50 \times 10^{-3} / 0.001 = 250$

λ $= 64/Re = 64/250 = 0.256$

Δp_a $= \lambda \, (l/d) \, (\varrho.v^2/2)$
$= 0.256 \times (50/0.05) \times (820 \times 5^2/2)$
$=26.24$ bar

Frictional Losses in Turbulent Flow

The inside surface of a round pipe is given in Figure 5.1. The mean height of the roughness is designated as 'ε', and the pipe inside diameter is designated as 'D'.

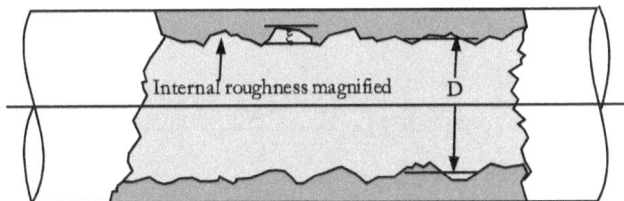

Internal roughness magnified

Figure 5.1 Relative roughness in a pipe

Relative roughness of the pipe's inside surface is defined as the mean roughness (ε) divided by the pipe inside diameter (D). That is,

Relative roughness = ε/D

Pipe roughness values depend on the pipe material and the manufacturing method.

Typical values of absolute roughness:

Drawn tubing – 0.0015 mm
Cast iron   – 0.26 mm
Riveted steel – 1.8 mm

Turbulent Flow, Smooth Pipes

In most fluid power systems, the pipes and hoses have smooth interiors, and the friction factor (λ) for smooth pipes can be calculated using the empirical formula given below.

$$\lambda = \frac{0.316}{\text{Re}^{0.25}}$$

The pressure loss (ΔPa) in a smooth pipe with the turbulent flow can now be calculated using the following formula:

$$\Delta\text{Pa} = 0.214 \frac{\mu^{0.25}\, l\, \varrho^{0.75}\, Q^{1.75}}{D^{4.75}}$$

Turbulent Flow, Rough Pipes

A close approximation of the friction factor for the turbulent flow through a pipe with rough interiors can be determined from the Swamee-Jain equation given below:

$$\lambda = \frac{0.25}{[\log_{10}(\varepsilon/3.7\,\text{d}) + \left(5.74/\text{Re}^{0.25}\right)]^2}$$

Summary of Equations (for Laminar Flow)

$$\text{Pressure loss, } \Delta\text{Pa} = \lambda \cdot \frac{l}{d} \cdot \frac{\varrho\, v^2}{2}$$

$$\text{Friction factor (laminar flow), } \lambda = \frac{64}{\text{Re}}$$

$$\text{Pressure loss (laminar flow), } \Delta\text{Pa} = \frac{32\, \mu\, l\, v}{d^2}$$

$$\text{Pressure loss (laminar flow), } \Delta\text{Pa} = \frac{128\, \mu\, l\, Q}{\pi\, d^4}$$

Example 5.5

Hydraulic 68-grade oil is flowing through a hydraulic line with an inside diameter of 0.0504 m at the rate of 0.0126 m³/s. Find the pressure drop for a 3.048 m length of hose. Assume the fluid density to be 880 Kg/m³.

Solution

Q	$= 0.0126 \text{ m}^3/\text{s}$
D	$= 0.0504 \text{ m}$
l	$= 3.048 \text{ m}$
ν	$= 6.8 \times 10^{-5} \text{ m}^2/\text{s}$
μ	$= \varrho\nu = 880 \times 6.8 \times 10^{-5}$
	$= 0.0598 \text{ Pa.s}$
V	$= 4Q/(\prod d^2) = 6.316 \text{ m/s}$
Re	$= V D / \nu$
	$= 6.316 \times 0.0504 / 6.8 \times 10^{-5}$
	$= 4681$

$$\Delta Pa = 0.214 \frac{\mu^{0.25} \, l \, \varrho^{0.75} \, Q^{1.75}}{D^{4.75}}$$

$$\Delta Pa = 0.214 \frac{0.0598^{0.25} \times 3.048 \times 880^{0.75} \times 0.0126^{1.75}}{0.0504^{4.75}}$$

$$= 36000 \text{ Pa}$$
$$= 0.36 \text{ bar}$$

Moody Diagram

The Moody Diagram, as given in Figure 5.2, plots the friction factor as a function of the Reynolds number and the relative pipe roughness. From the diagram, it can be inferred that for laminar flows, the friction factor depends only on the Reynolds number. For turbulent flows, the friction factor depends both on the Reynolds number and the relative roughness.

The following procedure can be followed to find the friction factor from the Moody Diagram:
- Find the value of Re
- Find the relative roughness ε
- Project a vertical line on the Re axis at the value of Re determined
- Find the curve corresponding to relative roughness
- Project horizontally to the f-axis to obtain the friction factor

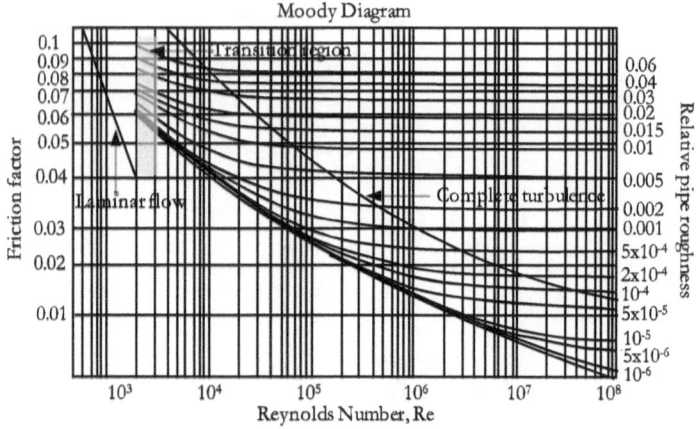

Figure 5.2 | Moody Diagram
Ref: Moody, L F (1944), 'Friction factors for pipe flow

Individual Pressure Losses

The individual pressure losses occur in pipe bends, junctions, and generally in pipe sections where the cross-sectional area or flow direction changes. The relation for the individual pressure losses is given by:

$$\Delta Pb = \zeta \, \frac{\varrho \cdot v^2}{2} = \zeta \, \frac{\varrho \cdot Q^2}{2\,A^2}$$

Where,

Δp_b = individual pressure loss [Pa]
ζ = individual resistance factor [-]
ϱ = the fluid density [kg/m3]
v = flow velocity [m/s]

Values of Loss Coefficient (ζ)

The value of the individual resistance factor ζ depends on the flow channel structure and dimensioning:

The loss coefficients (ζ):

- 90^0 elbow $- 0.2$
- 45^0 elbow $- 0.15$
- Tee fitting $- 0.9$
- Sharp-edged entrance $- 0.5$
- Rounded entrance $- 0.05$
- Sharp-edged exit $- 1.0$
- Rounded exit $- 1.0$

Individual pressure losses can also be found from the nomogram given in Figure 5.3

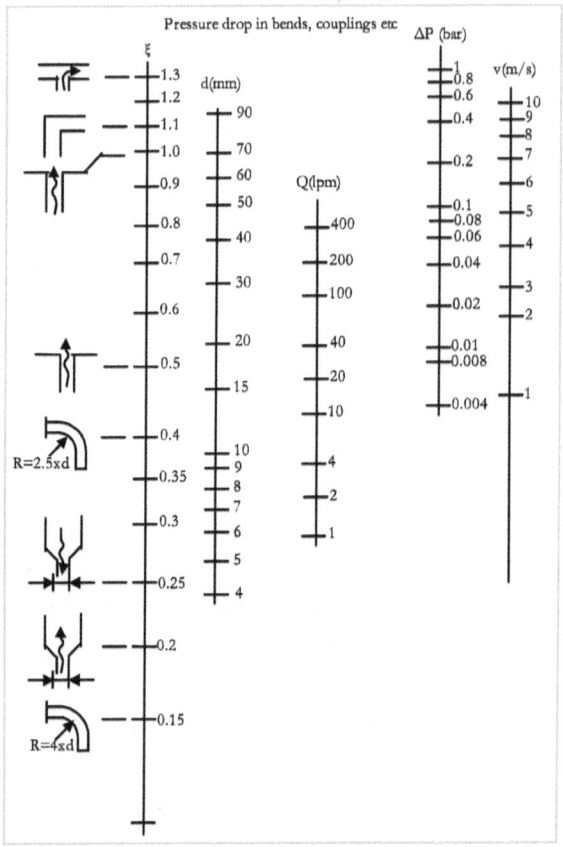

Figure 5.3 | Individual pressure losses

Total Pressure Losses

The total pressure loss in the piping is obtained by summing the frictional and individual pressure losses.

$$\Delta Ptot = \Delta Pa + \Delta Pb$$

The total pressure loss is limited to 3-5% for systems operating continuously. The total pressure loss is limited to 7-10% for systems with an intermittent duty cycle.

Chapter 6 | Installation, Routing & Maintenance of Fluid Conductors

The proper routing, installation, and maintenance of fluid conductors and their fittings are as essential as those of any other component in a hydraulic system. Installation stress, abrasion, tight bends, and exposure to higher pressures, temperatures, and corrosion reduce the service life of fluid conductors.

Conductor connections between components with relative motion should be made with hoses to prevent stresses, cracks, and vibrations in the connections and the associated fittings, such as couplings, flanges, or bolts.

One major problem with the fittings is the risk of leakage from loosening caused by shock and vibration.

The following sections examine the installation and maintenance of rigid hydraulic pipes, semi-rigid tubing, and flexible hoses, along with their fittings.

Installation of Hydraulic Conductors

It is essential to clean conductors and fittings before their installation. Remember to use the fewest fittings and connectors possible. The following bulleted lines give a few essential points for the proper installation of fluid conductors:

-Keep the length of conductors as small as possible to reduce pressure loss
-Avoid tight bends to minimize fluid turbulence and pressure loss

-Route the fluid conductors optimally to minimize the pressure loss and leakage in the system and reduce their abrasion, rubbing, kinking, and excessive flexing

-Restrain, support, protect, and guide fluid conductors using clamps at frequent intervals to prevent chafing against one another and minimize the vibration of the conductor system

-Avoid crossing two hose lines. However, when the crossing is unavoidable, join the two lines at the junction point

-Use proper tools for preparing a conductor for connecting to another conductor or a component

-Conductors need to be flushed thoroughly with a suitable degreasing agent immediately after their installation

Hose Assembly Routing Tips

It is better to avoid sharp bends when routing hoses. Using a bent or kinked hose causes severe backpressure in the associated system. It may also cause internal damage to the hose, leading to premature failure. Some tips for routing a hose assembly, especially about its length, minimum bend radii, and multi-plane bending, are given below:

-The hose length must be slightly longer than the actual distance between two linear connections to accommodate the changes in length with pressure changes

-Ensure that the hose is slack on both sides of the clamp to compensate for contraction and expansion

-The bend radius of the hose must be as large as possible to avoid the hose collapsing or the flow restriction

-As far as possible, bend a hose in one plane only. This precaution prevents the wire reinforcement from twisting and improves its pressure capacity. The multi-plane bending of a piece of the hose can often be avoided by rerouting the hose

-If the multi-plane bending cannot be avoided, install a clamp between the bends and provide enough hose length on both sides of the clamp to relieve the strain on the hose's reinforcement wires

-Use clamps to secure the length of the hose in position and to keep it from rubbing against adjacent surfaces

-The hose connected to a cylinder that undergoes a pivoting motion must be of proper length to avoid its kinking or bending beyond its minimum bend radius

-Use appropriate swivel joints for the hose connection to reduce the bending transmitted to the hose assembly by the relative motion between the associated machine elements

-The use of carriers keeps the hoses neatly nestled to prevent their rubbing against each other

How to Replace a Hydraulic Hose?

Hydraulic hoses deteriorate over time, which can cause leaks. Locate the damaged hose and remove it from the equipment. Before starting the replacement process, it is important to take some safety precautions. Wear goggles and gloves to protect eyes and hands. Also, it is important to ensure that all trapped pressure in the system is relieved to prevent accidents. Place a bucket under the hose to catch any leaking fluid. Then, wash off the connectors to prevent debris from entering the system. Finally, wrenches are used to unscrew the fittings, remove the damaged hose, and install a matching replacement.

Installing a New Hose

To install a new hose, ensure it matches the old one's specifications. Then slide an abrasion sleeve over the hose and wipe off any dirt from the hose fittings. Screw one end of the hose into the fitting and tighten it to the correct torque. Finally, insert the other end of the hose into the connector and tighten it securely with two wrenches.

Maintenance of Hydraulic Conductors

Hydraulic conductors and fittings should have the correct pressure ratings (wall thickness and material). When using semi-rigid conductors, seamless precision steel tubes are used. The following points generalize the maintenance activities of hydraulic conductors and fittings.

-Provide conductors with enough support throughout their entire length
-Inspect the conductors and their routing for damages, defects, and displacement
-Examine the conductors and their joints for leakage, looseness, scratches, kinks, and burrs
-Tighten any loose fittings or nut connections with the correct amount of torque to minimize leakage and reduce contamination
-Repair or replace defective conductors or fittings
-Ensure that any pipe replacement is of the same length, size, and wall thickness

Fittings and Joints

Pipes use thread joints, either tapered or straight, to produce a leak-proof metal-to-metal seal. BSP threaded connections on components allow the application of sealed edge or O-ring seal connectors.

Flaring, brazing, or couplings can quickly and easily join tubes. Flared or flareless fittings are used for tubing end connections.

Hose fittings can be either permanent or reusable. Permanent hose fittings are installed on the hose by crimping and cannot be disassembled. Quick-disconnect couplings are used for convenience because they can be installed and removed by hand and are useful when repeated line connections and disconnections are required.

Leak Detection

Fluorescent dyes can be mixed with the fluid to detect leaks. A blend of two fluorescent dyes - one fluoresces best under ultraviolet and the other under blue light - can be combined with the fluid. When scanned with ultraviolet or blue light, the dye glows bright yellow-green.

Case Study #6.1 | Maintaining the Conductor System

A professional team is responsible for maintaining a plant hydraulic system, including pipes, tubes, hoses, and fittings. List some points for maintaining the conductor system.

Answer

- Select the correct sizes and materials for the conductors and fittings.
- Use a tube cutter to cut small seamless tubes.
- Use a machine equipped with a grinding disc to cut larger pipes and tubes.
- Do not cut pipes using a flame-cutting machine.
- Clean all welded joints inside the pipe thoroughly by grinding.
- Clean pipe bores using a rotary steel brush.
- Do not sandblast internal pipe surfaces.
- Close all pipe ends with plastic caps to maintain cleanliness before assembly.

Case Study #6.2 | Steps to Clean Pipe Sections

Pipe sections in a hydraulic system for a marine application must be cleaned before assembly. Enumerate the typical steps for cleaning pipe sections.

Answer

- Submerge pipe sections in an acid bath for 2 to 24 hours, depending on their cleanliness. The acid bath should typically contain 15% hydraulic fluid.
- Wash pipe sections thoroughly with fresh water.
- Neutralize pipe sections by immersing them in a weak alkaline bath solution.
- Dry the pipe sections with warm air
- Apply a coat of protective oil film

Flushing of Pipelines

It is crucial to ensure that all components, including fluid conductors and fittings, are supplied clean and ready for assembly into the system, with all openings sealed. The piping should be free from scale, rust, and flux.

Once all the components are assembled, the system should be flushed to remove all foreign particles. Flushing can also be performed regularly to remove contaminants from hydraulic fluid and improve system performance.

After flushing, the system should be started following the proper procedure.

Flushing hydraulic pipelines is a meticulous process that involves removing particles, wear, dust, debris, and welding slag from the pipelines. This is achieved by passing hot oil through the pipes at a high speed. The temperature and flow rate of the flushing liquid are designed to thoroughly dislodge contaminants, forcing them through the system and flushing them out, ensuring a clean, efficient pipeline.

Flushing helps the system remove contaminants and maintain the required cleanliness of the fluid. The absence of flushing can lead to rapid component wear, system malfunction, and eventual system breakdown.

A flushing unit typically comprises a reservoir, higher-capacity pumps, a pressure relief valve, a filter with a clogging indicator, and hose connections. The filter element should have a generous dirt-holding capacity. Normal system oil can be used during flushing.

During flushing, bypass system components with flow restrictions or that may be damaged by the high flushing flow. Failure to do so can result in system damage or even failure.

In some cases, filters may be left in the system with the filter element removed during flushing to prevent damage.

A large system should be divided into sections, and each section, including pump stations and valve manifolds, should be flushed separately.

The flushing velocity and temperature should be maximized.

A reasonable temperature for mineral fluids is 60°C (140°F).
The fluid velocity should preferably be at least twice the system's rated velocity and, in any case, at least 8 m/s (26 ft/s).

The Reynolds number should exceed 3000 to ensure turbulent flow.

A Typical Flushing Procedure

With the system prepared, circulate the flushing fluid at the required flow rate and temperature, typically for two to three hours, until the pipelines are clean. Some proactive methods for cleaning pipelines are highlighted below.

Contaminants and insoluble suspensions can be removed by filtration at normal flow rates. High-turbulence flushing can enhance flushing by lowering the flushing fluid viscosity and increasing fluid flow rates using specialized equipment.

After flushing, all fluid must be drained as hot and as quickly as possible. Then, replace the filters and clean the reservoir again.

Start-up Procedures

Below is the startup procedure for filling the fluid and priming the pump during initial setup or a regular fluid changeover.

Fill the reservoir to approximately 75% with the selected fluid. Run the pump for 15 seconds with the PRV or bleed valve wide open, then stop and let it sit for 45 seconds. Repeat this procedure a few times to prime the pump.

Run the pump for a minute with the bypass or pressure relief open. Stop the pump and let it sit for a minute.

Close the bypass and allow the pump to operate under load without actuator operation for no more than five minutes. Stop the pump and let the system sit for about five minutes.

Start the pump and operate the actuators one at a time, allowing fluid to return to the reservoir before moving to the next actuator. After operating the final actuator, shut down the system. Monitor the fluid level in the reservoir.

If the level drops below 25%, refill the reservoir to 75% and run the system in five-minute intervals. At each shutdown, bleed the air from the system. Pay close attention to the system sounds to determine if the pump is cavitating.

Run the system for 30 minutes to reach normal operating temperature. Shut down the system and replace the filters. Inspect the reservoir for obvious signs of cross-contamination. If any indication of cross-contamination is present, drain and flush the system again.

After six hours of operation, shut down the system, replace the filter, and test the fluid. The system must be monitored for a while to ensure that it is fully flushed and purged of the old fluid before introducing the new fluid.

Summary of Points for the Installation and Maintenance of Conductors

Figure 6.1 summarizes the points for the installation, routing, and maintenance of fluid conductors.

Installation, Routing, and Maintenance of Fluid Conductors

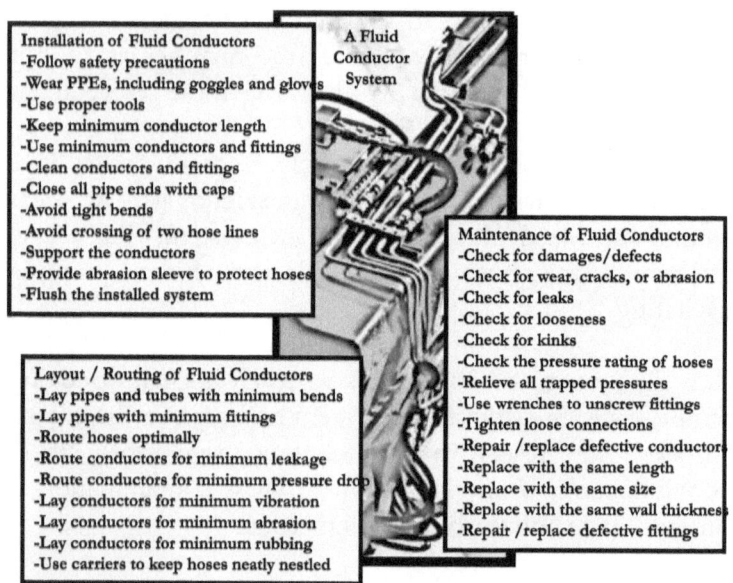

Installation of Fluid Conductors
-Follow safety precautions
-Wear PPEs, including goggles and gloves
-Use proper tools
-Keep minimum conductor length
-Use minimum conductors and fittings
-Clean conductors and fittings
-Close all pipe ends with caps
-Avoid tight bends
-Avoid crossing of two hose lines
-Support the conductors
-Provide abrasion sleeve to protect hoses
-Flush the installed system

A Fluid Conductor System

Maintenance of Fluid Conductors
-Check for damages/defects
-Check for wear, cracks, or abrasion
-Check for leaks
-Check for looseness
-Check for kinks
-Check the pressure rating of hoses
-Relieve all trapped pressures
-Use wrenches to unscrew fittings
-Tighten loose connections
-Repair /replace defective conductor
-Replace with the same length
-Replace with the same size
-Replace with the same wall thickness
-Repair /replace defective fittings

Layout / Routing of Fluid Conductors
-Lay pipes and tubes with minimum bends
-Lay pipes with minimum fittings
-Route hoses optimally
-Route conductors for minimum leakage
-Route conductors for minimum pressure drop
-Lay conductors for minimum vibration
-Lay conductors for minimum abrasion
-Lay conductors for minimum rubbing
-Use carriers to keep hoses neatly nestled

Figure 6.1 | Maintenance of fluid conductors

Reference: This chapter is taken from the textbook 'Maintenance, Troubleshooting, and Safety in Hydraulic Systems' by the same author, Joji Parambath. Maintenance of other hydraulic components is described in the referenced book.

Fluid Conductor Faults

Table 6.1 lists faults in fluid conductors and their remedial actions.

Table 6.1 | Fluid conductor faults

Fault	Remedy
Weakened tubing or hose	-Use a compatible fluid
Hose tube cracked	-Protect hoses from excessive heat
Hose bursts	-Use the correct hose to withstand high-pressure surges -Alter environmental or operating conditions -Use a correct length hose
Excessive pressure drop in the hose	-Use a hose of the correct size -Improve bore condition
Faulty fittings	-Replace faulty ones
Leakage through fittings	-Replace damaged seal -Correct damaged threads -Use clamps to prevent loosening of fittings

7 | Objective Type Questions

1. The schedule number is associated with:
a) The wall thickness of a pipe
b) Viscosity index
c) The acidity level of a hydraulic fluid
d) The hardness of a seal material

2. Mark the least expensive hydraulic fluid conductor system?
a) Pipes
b) Tubing
c) Hoses
d) Cannot be distinguished

3. The dash number is used to specify:
a) Inside diameter (ID) of a hose
b) Wall thickness of the tubing
c) The elasticity of seal materials
d) Contamination concentration level

4. Mark the <u>incorrect</u> statement
a) Hoop stress of a given length of pipe is the circumferential stress acting on the wall of the pipe under the operating pressure.
b) Bending hose/tubing to a radius smaller than its rated minimum bend radius may cause premature failure.
c) Flared or flareless fittings can be used for tubing end connections.
d) Hose length must be precisely equal to the actual distance between their end connections.

8 | Review Questions

1) What is the function of fluid conductors in hydraulic systems?

2) List three primary types of fluid conductors used in hydraulic systems. Briefly explain their degree of flexibility.

3) What are the basic requirements for the satisfactory function of the fluid conductor system in a hydraulic circuit?

4) State the reason for the probable energy loss in the fluid conductors used in hydraulic systems.

5) What are the reasons for the fluid leakages in a hydraulic distribution system?

6) What are the ways to minimise fluid leakage in a hydraulic fluid distribution system?

7) What is the definition of 'schedule number' when referring to piping and fittings in hydraulic systems?

8) What variables determine the wall thickness and the factor of safety of a fluid conductor?

9) Why should fluid conductors have greater strength than the system working pressure requires?

10) Briefly, explain the terms of fluid conductors: (1) Bend radius, (2) Tensile stress, (3) Burst pressure, and (4) Working pressure.

11) What factors determine the pressure rating of a fluid conductor?

12) List out the procedure to calculate the size of a fluid conductor for a hydraulic system.

13) Explain the purpose of hydraulic pipes, briefly.

14) State two disadvantages of using pipes in hydraulic systems

15) How are the pipes used as fluid conductors in hydraulic systems specified?

16) Explain how the wall thickness of the pipe used as a fluid conductor in a hydraulic system is specified.

17) What is meant by the schedule number of a standard pipe, as used in hydraulic systems?

18) How is the pipe size classified in hydraulic systems?

19) Describe the methods of coupling hydraulic pipes.

20) What are the functions of hydraulic pipe threads?

21) State some common materials used in the manufacturing of hydraulic pipes.

22) State the major disadvantages of steel pipes, as used in hydraulic systems.

23) What are the two types of thread configurations used in the hydraulic piping systems? Differentiate them.

24) Name two common types of hydraulic pipe joints.

25) What is the disadvantage of hydraulic threaded fittings?

26) Briefly explain the use of tubing in hydraulic systems.

27) Why is steel tubing more commonly used than steel pipe in hydraulic systems?

28) State the common materials used for manufacturing hydraulic tubing.

29) How do you specify hydraulic tubing?

30) What factors determine the tubing size?

31) Distinguish between the thin-walled and the thick-walled hydraulic conductors.

32) Mention one advantage and one disadvantage of hydraulic tubing.

33) Mention two advantages of the hydraulic tubing over the pipes

34) Describe any two methods of coupling the tubing in hydraulic systems.

35) Describe different types of tube fittings, as used in hydraulic systems.

36) Briefly explain the correct methods for bending and flaring hydraulic tubing.

37) What is a flare fitting, as used in hydraulic systems?

38) Why is flaring needed, and how is it done?

39) What is the difference between the flared fitting and the compression fitting, as used in hydraulic systems?

40) List the parts of a flared tubing fitting assembly, as used in hydraulic systems.

41) List the parts of a flare-less tubing fitting assembly, as used in a hydraulic system.

42) Briefly explain the use of hoses in hydraulic conductor systems.

43) List three essential elements of a flexible hose used in hydraulic systems.

44) Describe the basic constructional features of flexible hoses, as used in hydraulic systems?

45) What is the purpose of providing a protective outer layer for a hydraulic hose?

46) Under what conditions are flexible hoses used in hydraulic systems?

47) How are the hoses used as fluid conductors in hydraulic systems specificd?

48) What does the dash number of a hydraulic hose refer to?

49) What determines the pressure rating of a hydraulic hose?

50) Explain how the pressure rating of hydraulic hoses is increased.

51) What are the advantages of hydraulic hoses?

52) Mention three factors that are to be considered while selecting hydraulic hoses.

53) Briefly explain five essential parameters to be considered while selecting a hydraulic hose.

54) Give a brief note on (1) Hose motion in hydraulic systems and (2) The applications of hydraulic hoses.

55) Explain the purpose of the quick disconnect coupling, as used in hydraulic systems.

9 | Numerical Problems

1) Calculate the minimum inside diameter of the hydraulic suction pipe to handle the flow rate of 40 lpm with an average fluid velocity not exceeding 0.5 m/s. [Ans: 4.12 cm]

2) Determine the size of the pressure line of a hydraulic system with a 37.85 lpm positive-displacement pump. The recommended flow velocity through the pressure line is 5 m/s. [Ans: 12.7 mm]

3) Determine the size of the return line of a hydraulic system with a 0.00103 m^3/s positive-displacement pump. The recommended return flow velocity is 2.5 m/s. [Ans: 23 mm]

4) A hydraulic system is to permit the flow rate of 40 lpm with an average fluid velocity not exceeding 4 m/s. Calculate the minimum inside diameter of the pressure conductor in the system. [Ans: 1.45 cm]

5) A hydraulic system is to permit the flow rate of 40 lpm with an average fluid velocity not exceeding 2 m/s. Calculate the minimum inside diameter of a return-line conductor. [Ans: 2.06 cm]

6) Find the schedule number of a steel pipe for a hydraulic system at the estimated working pressure of 112.5 bar. The allowable stress is 4000 bar.[Ans: 30]

7) Calculate the burst pressure of a seamless cold-drawn steel tubing of outside diameter 25 mm and wall thickness 2.5 mm. The tubing has a tensile strength of 400 MPa. [Ans: 88.88 MPa]

8) Determine the safe working pressure for a steel tube with a burst pressure of 900 bar, assuming a safety factor of 8. [Ans: 112.5 bar]

9) A hydraulic pipe has an outer diameter of half an inch and schedule 40, and another pipe has the same outer diameter but of schedule 80. State the difference between these two pipes, explicitly.

10 | References

1) Anthony Esposito, Fluid Power with Applications, 6th Edition, Prentice-Hall of India, 2006

2) Article on: 'About Hydraulic Hose', GlobalSpec Inc., Jordan Rd, Troy, NY, USA

3) Article on: 'Hose and tubing assemblies, Hydraulic hose, Hose installation', Hydraulics & Pneumatics Magazine, The Penton Media Building, Cleveland, OH, USA

4) Article on: 'Hydraulic system tubing – Lifelines to power and motion control', by Terry Karl and Mark Morrow

5) Catalogue on 'High-Pressure Stainless Steel Hoses' PARKER / PAGE International Hose, Texas, www.pageintl.com

6) Catalogue on 'Hose and Flexible Tubing', Document No. R8 MS-01-180, Swagelok Company, U.S.A., AGS October 2013

7) Catalogue on: 'Hose, Fittings, and adapters catalogue 2010' by Alfagomma Hydraulic spa, Vimercate, MI, Italy, http://www.alfagomma.com/

8) Paper on: 'Dash Number Chart', Jones Enterprise, LaPorte, Indiana, USA

9) Document on: 'Flexible metal hoses', HAM-LET Advanced Control Technology, info@ham-let.com

10) Document on: 'How We Bend Steel Tubing and Steel Beams', Paramount Roll and Forming Inc, Los Angeles, California

11) Document on: 'Hydraulic Hose Installation', Airline Hydraulics Corporation, Bensalem, PA, USA

12) Document on: 'PROPER INSTALLATION & HOSE ROUTING', Good Year, www.hydraulics.goodyear.com

13) Documents on: 'Introduction to Hydraulics', 'Hydraulic Hose Life Made Simple', GATES Professional Development series, The Gates Rubber Company, Colorado, USA

Fluid Power Educational Series Books

1. Pneumatic Systems and Circuits -Basic Level (In the SI Units)
2. Industrial Pneumatics -Basic Level (In the English Units)
3. Pneumatic Systems and Circuits -Advanced Level
4. Electro-Pneumatics and Automation
5. Design of Pneumatic Systems (In the SI Units)
6. Design Concepts in Pneumatic Systems (In the English Units)
7. Maintenance, Troubleshooting, and Safety in Pneumatic Systems
8. Industrial Hydraulic Systems and Circuits -Basic Level (In the SI Units)
9. Industrial Hydraulics -Basic Level (In the English Units)
10. Hydraulic Fluids
11. Hydraulic Filters: Construction, Installation Locations, and Specifications
12. Hydraulic Power Packs (In the SI Units)
13. Power Packs in Hydraulic Systems (In the English Units)
14. Hydraulic Cylinders (In the SI Units)
15. Hydraulic Linear Actuators (In the English Units)
16. Hydraulic Motors (In the SI Units)
17. Hydraulic Rotary Actuators (In the English Units)
18. Hydraulic Accumulators and Circuits (In the SI Units)
19. Accumulators in Hydraulic Systems (In the English Units)
20. Hydraulic Pipes, Tubes, and Hoses (In the SI Units)
21. Pipes, Tubes, and Hoses in Hydraulic Systems (In the English Units)
22. Design of Industrial Hydraulic Systems (In the SI Units)
23. Design Concepts in Industrial Hydraulic Systems (In the English Units)

24. Maintenance, Troubleshooting, and Safety in Hydraulic Systems

25. Hydrostatic Transmissions (HSTs) (In the SI Units)

26. Concepts of Hydrostatic Transmissions (In the English Units)

27. Load Sensing Hydraulic Systems (In the SI Units)

28. Concepts of Load Sensing Hydraulic Systems (In the English Units)

29. Electro-hydraulic Proportional Valves

30. Electro-hydraulic Servo Valves

31. Cartridge Valves

32. Electro-hydraulic Systems and Relay Circuits

33. Practical Book: Pneumatics - Basic Level

34. Practical Book: Electro-pneumatics - Basic Level

35. Practical Book: Industrial Hydraulics – Basic Level

36. Programmable Logic Controllers and Programming Concepts

37. Compressed Air Dryers

38. Hydraulic Circuits – Identification of Components and Analysis

For more details, please visit: **https://jojibooks.com.**

About the Author

Joji Parambath is an accomplished expert in Pneumatics, Hydraulics, and PLCs with 25 years of experience. He has trained many professionals from diverse industries, faculty members, and engineering students throughout his career.

Joji is the primary faculty member at Fluidsys Training Centre in Bangalore, India, offering comprehensive training in Pneumatics and Hydraulics. He has authored 39 books on pneumatics and hydraulics, all designed to convey the subject in a simple, easy-to-understand manner.

Joji attributes the creation of his book series to the active engagement and valuable suggestions of his trainees during the training programs. He would like to extend his gratitude towards them.

10th June 2020